THE
SCEPTICAL CHYMIST

The Classic 1661 Text

Robert Boyle

DOVER PUBLICATIONS, INC.
Mineola, New York

Copyright

Copyright © 2003 by Dover Publications, Inc.
All rights reserved.

Bibliographical Note

This Dover publication, first published in 2003, is an unabridged republication of the work originally published in 1911 by J. M. Dent and Sons, Ltd., London. The Publisher's Note, page v, is newly added.

Library of Congress Cataloging-in-Publication Data

Boyle, Robert, 1627–1691.
    The sceptical chymist / Robert Boyle.
       p. cm.
    Originally published: London : J. M. Dent & Sons, Ltd.; New York : E. P. Dutton, 1911.
    ISBN 0-486-42825-7 (pbk.)
    1. Chemistry—Early works to 1800. I. Title.

QD31.3.B69 2003
540—dc21

                                   2002041347

Manufactured in the United States of America
Dover Publications, Inc., 31 East 2nd Street, Mineola, N.Y. 11501

NOTE

Robert Boyle, whose life spanned the seventeenth century, is best remembered for "Boyle's Law," explaining the relationship between the pressure and the volume of a gas. A radical thinker from the start, the young Irishman joined England's so-called "Invisible College" whose aim was to cultivate concepts termed the "new philosophy." Reflecting a point of view that gripped Boyle throughout his career, this group sought out new methods of experimental science to prove or disprove scientific hypotheses through controlled experiments—a foundation of modern scientific thought. So influential was the College that it was granted a kingly charter allowing it to become the Royal Society of London for Improving Natural Knowledge. (A member of its first council, Boyle was later elected president of the Society, but declined the honor.)

Boyle's active research in Oxford, dealing largely with the behavior of gases including the earth's atmosphere, led to his first book, *The Spring of Air* (1660). Barely a year later, he published *The Sceptical Chymist,* criticizing earlier research founded on the belief that salt, mercury, and sulphur were the "true principles of things," and setting forward a new view that matter's basic elements consisted of various sorts and sizes of particles—"corpuscles," he called them—capable of arranging themselves into groups, each a chemical substance.

The first to coin the term *analysis*. . . to determine a substance as an acid or base . . . the first to study the emission of light from living organisms, and to study the phenomena of oxygen and combustion even before they were properly understood . . . the first to use a hydrometer to measure a liquid's density . . . and (hard to believe) the man credited with the invention of the sulphur-tipped match, Robert Boyle is rightfully honored today as a founder of modern chemistry and a pioneer in the acceptance of the scientific method.

His own words, from *The Sceptical Chymist,* reflect the freshness of his thinking and his view of a world yet to be explored:

" . . . methinks the chymists, in their searches after truth, are not unlike the navigators of Solomon's Tarshish fleet, who brought home from their long and tedious voyages, not only gold, and silver, and ivory, but apes and peacocks too; for so the writings of several of your hermetick philosophers present us, together with divers substantial and noble experiments, theories, which either like peacocks' feathers make a great shew, but are neither solid nor useful; or else like apes, if they have some appearance of being rational, are blemished with some absurdity or other, that when they are attentively considered, make them appear ridiculous."

CONTENTS

INTRODUCTORY PREFACE TO THE FOLLOWING TREATISE

To give the reader an account, why the following treatise is suffered to pass abroad so maimed and imperfect, I must inform him that 'tis now long since, that to gratify an ingenious gentleman, I set down some of the reasons that kept me from fully acquiescing either in the peripatetical, or in the chymical doctrine, of the material principles of mixt bodies. This discourse some years after falling into the hands of some learned men, had the good luck to be so favourably received and advantagiously spoken of by them, that having had more than ordinary invitations given me to make it public, I thought fit to review it, that I might retrench some things that seemed not so fit to be shewn to every reader, and substitute some of those other things that occurred to me of the trials and observations I had since made: What became of my papers, I elsewhere mention in a Preface where I complain of it: but since I writ that, I found many sheets that belonged to the subjects I am now about to discourse of. Wherefore seeing that I had then in my hands as much of the first dialogue as was requisite to state the case, and serve for an introduction as well to the conference betwixt Carneades and Eleutherius, as to some other dialogues, which for certain reasons are not herewith published, I resolved to supply, as well as I could, the contents of a paper belonging to the second of the following discourses, which I could not possibly retrieve, though it were the chief of them all. And having once more tried the opinion of friends, but not the same, about this imperfect work, I found it such, that I was content in compliance with their desires, that not only it should be published, but that it should be published as soon as conveniently might be. I had indeed all along the dialogues spoken of myself as

of a third person; for they containing discourses which
were among the first treatises that I ventured long ago
to write of matters philosophical, I had reason to desire,
with the painter, to *latere pone tabulam,* and hear what
men would say of them, before I owned myself to be their
author. But besides that now I find, 'tis not unknown to
many who it is that writ them, I am made to believe that
'tis not inexpedient they should be known to come from
a person altogether a stranger to chymical affairs. And
I made the less scruple to let them come abroad uncom-
pleated, partly because my affairs and pre-ingagements
to publish divers other treatises allowed me small hopes of
being able in a great while to compleate those dialogues,
and partly because I am not unapt to think, that they may
come abroad seasonably enough, though not for the
author's reputation, yet for other purposes. For I observe,
that of late chymistry begins, as indeed it deserves, to be
cultivated by learned men who before despised it; and
to be pretended to by many who never cultivated it, that
they may be thought not to be ignorant of it: whence it is
come to pass, that divers chymical notions about matters
philosophical are taken for granted and employed, and
so adopted by very eminent writers both naturalists
and physicians. Now this I fear may prove somewhat
prejudicial to the advancement of solid philosophy: for
though I am a great lover of chymical experiments, and
though I have no mean esteem of divers chymical remedies,
yet I distinguish these from their notions about the causes
of things and their manner of generation. And for ought
I can hitherto discern, there are a thousand phænomena in
nature, besides a multitude of accidents relating to the
human body, which will scarcely be clearly and satisfactorily made out by them that confine themselves to
deduce things from salt, sulphur, and mercury, and the
other notions peculiar to the chymists, without taking
much more notice than they are wont to do, of the motions
and figures, of the small parts of matter and the other
more catholic and fruitful affections of bodies. Wherefore it will not perhaps be now unseasonable to let our
Carneades warne men, not to subscribe to the grand doctrine

of the chymists touching their three hypostatical principles, till they have a little examined it, and considered
how they can clear it from his objections, divers of
which 'tis like they may never have thought on; since
a chymist scarce would, and none but a chymist could
propose them. I hope also it will not be unaccept-
able to several ingenious persons, who are unwilling to
determine of any important controversie, without a
previous consideration of what may be said on both sides,
and yet have greater desires to understand chymical
matters than opportunities of learning them, to find here
together, besides several experiments of my own purposely made to illustrate the doctrine of the elements,
divers others scarce to be met with, otherwise then
scattered among many chymical books: and to find
these associated experiments so delivered as that an
ordinary reader, if he be but acquainted with the usual
chymical termes, may easily enough understand them;
and even a wary one may safely rely on them. These
things I add, because a person anything versed in the
writings of chymists cannot but discern by their obscure,
ambiguous, and almost ænigmatical way of expressing
what they pretend to teach, that they have no mind to be
understood at all, but by *the sons of Art* (as they call them),
nor to be understood even by these without difficulty and
hazardous trials. Insomuch that some of them scarce
ever speak so candidly, as when they make use of that
known chymical sentence: *Ubi palam locuti fumus, ibi
nihil diximus.* And as the obscurity of what some writers
deliver makes it very difficult to be understood; so the
unfaithfulness of too many others makes it unfit to be
relied upon. For though unwillingly, yet I must for the
truth sake, and the reader's, warne him not to be forward
to believe chymical experiments when they are set down
only by way of prescriptions, and not of relations; that is,
unless he that delivers them mentions his doing it upon
his own particular knowledge, or upon the relation of
some credible person, avowing it upon his own experience. For I am troubled, I must complain, that even
eminent writers, both physitians and philosophers, whom

I can easily name, if it be required, have of late suffered themselves to be so far imposed upon, as to publish and build upon chymical experiments, which questionless they never tried; for if they had, they would, as well as I, have found them not to be true. And indeed it were to be wished, that now that those begin to quote chymical experiments that are not themselves acquainted with chymical operations, men would leave off that indefinite way of vouching the chymists say this, or the chymists affirm that, and would rather for each experiment they alleged name the author or authors upon whose credit they relate it; for, by this means they would secure themselves from the suspicion of falsehood (to which the other practice exposes them), and they would leave the reader to judge of what is fit for him to believe of what is delivered, whilst they employ not their own great names to countenance doubtful relations; and they will also do justice to the inventors or publishers of the true experiments, as well as upon the obtruders of false ones. Whereas by that general way of quoting the chymists, the candid writer is defrauded of the particular praise, and the impostor escapes the personal disgrace that is due to him.

The remaining part of this Preface must be imployed in saying something for Carneades, and something for myself.

And first, Carneades hopes that he will be thought to have disputed civilly and modestly enough for one that was to play the antagonist and the sceptic. And if he anywhere seem to slight his adversaries tenents and arguments, he is willing to have it looked upon as what he was induced to, not so much by his opinion of them, as the examples of Themistius and Philoponus, and the custom of such kind of disputes.

Next, in case that some of his arguments shall not be thought of the most cogent sort that may be, he hopes it will be considered that it ought not to be expected that they should be so. For, his part being chiefly but to propose doubts and scruples, he does enough, if he shews that his adversaries arguments are not strongly concluding, though his own be not so neither. And if there should appear any disagreement betwixt the things he delivers in divers passages, he hopes it will be considered, that it is not necessary that all the things a sceptic proposes should be consonant; since it being his work to suggest doubts against the opinion he questions, it is allowable for him to propose two or more several hypotheses about the same thing: and to say that it may be accounted for this way, or that way, or the other way, though these wayes be perhaps inconsistent among themselves. Because it is enough for him, if either of the proposed hypotheses be but as probable as that he calls in question. And if he propose many that are each of them probable, he does the more ratify his doubts, by making it appear the more difficult to be sure, that that way which they all differ from is the true. And our Carneades by holding the negative, has this advantage, that if among all the instances he brings to invalidate the vulgar doctrine of those he disputes with, any one be irrefragable, that alone is sufficient to overthrow a doctrine which universally asserts what he opposes. For, it cannot be true, that all bodies whatsoever that are reckoned among the perfectly mixt ones, are compounded of such a determinate number of such or such ingredients, in case any one such body can be produced that is not so compounded; and he hopes too, that accurateness will be the less expected from him, because his undertaking obliges him to maintain such opinions in chymistry, and that chiefly by chymical arguments, as are contrary to the very principles of the chymists, from whose writings it is not therefore like he should receive any intentional assistance, except from some passages of the bold and ingenious Helmont, with whom he yet disagrees in many things (which reduce him to explicate divers chymical phænomena, according to other notions): and of whose ratiocinations, not only some seem very extravagant, but even the rest are not wont to be as considerable as his experiments. And though it be true indeed, that some Aristotelians have occasionally written against the chymical doctrine he oppugnes, yet since they have done it according to their principles, and since our

Carneades must as well oppose their hypothesis as that
of the spagyrist, he was fain to fight his adversaries with
his own weapons, those of the peripatetic being improper
if not hurtful for a person of his tenets; besides that
those Aristotelians (at least those he met with), that have
written against the chymists, seem to have had so little
experimental knowledge in chymical matters, that by
their frequent mistakes and unskilful way of oppugning,
they have too often exposed themselves to the derision of
their adversaries, for writing so confidently against what
they appeare so little to understand.

And lastly, Carneades hopes he shall do the ingenious
this piece of service, that by having thus drawn the
chymists' doctrine out of their dark and smokie laboratories, and both brought it into the open light, and shewn
the weakness of their proofs, that have hitherto been wont
to be brought for it, either judicious men shall henceforth
be allowed calmly and after due information to disbelieve
it, or those abler chymists, that are zealous for the reputation of it, will be obliged to speak plainer than hitherto
has been done, and maintain it by better experiments and
arguments than those Carneades hath examined: so that
he hopes the curious will one way or other derive either
satisfaction or instruction from his endeavours.   And as
he is ready to make good the profession he makes in the
close of his discourse, of being ready to be better informed,
so he expects either to be indeed informed, or to be let
alone.   For though, if any truly knowing chymists shall
think fit in a civil and rational way to shew him any truth
touching the matter in dispute that he yet discernes not,
Carneades will not refuse either to admit, or to own a
conviction: yet if any impertinent person shall, either to
get himselfe a name, or for what other end soever, wilfully
or carelessly mistake the state of the controversie, or the
sense of his arguments, or shall rail instead of arguing, as
hath been done of late in print by divers chymists; or
lastly, shall write against them in a canting way, I mean
shall express himselfe in ambiguous or obscure termes, or
argue from experiments, not intelligibly enough delivered,
Carneades professes that he values his time so much, as

not to think the answering such trifles worth the loss of it.

And now having said thus much for Carneades, I hope the reader will give me leave to say something for myself.

And first, if some morose readers shall find fault with my having made the interlocutors upon occasion complement with one another, and that I have almost all along written these dialogues in a style more fashionable than that of mere scholars is wont to be, I hope I shall be excused by them that shall consider, that to keep a due decorum in the discourses it was fit that in a book written by a gentleman, and wherein only gentlemen are introduced as speakers, the language should be more smooth and the expressions more civil than is usual in the more scholastic way of writing. And indeed, I am not sorry to have this opportunity of giving an example how to manage even disputes with civility; whence perhaps some readers will be assisted to discern a difference betwixt bluntness of speech and strength of reason, and find that a man may be a champion for truth without being an enemy to civility; and may confute an opinion without railing at them that hold it; to whom he that desires to convince and not to provoke them, must make some amends by his civility to their persons, for his severity to their mistakes; and must say as little else as he can to displease them, when he says that they are in an error.

But perhaps other readers will be less apt to find fault with the civility of my disputants than the chymists will be, upon the reading of some passages of the following dialogue, to accuse Carneades of asperity. But if I have made my sceptic sometimes speak slightingly of the opinions he opposes, I hope it will not be found that I have done any more than became the part he was to act of an opponent: especially if what I have made him say be compared with what the prince of the Romane orators himself makes both great persons and friends say of one another's opinions, in his excellent dialogues, *De Natura Deorum:* and I shall scarce be suspected of partiality in the case, by them that take notice that there is full as much (if not far more) liberty of slighting their adversaries tenets

to be met with in the discourses of those with whom
Carneades disputes. Nor need I make the interlocutors
speak otherwise than freely in a dialogue, wherein it was
sufficiently intimated that I meant not to declare my own
opinion of the arguments proposed, much lesse of the
whole controversy itselfe, otherwise than as it may by an
attentive reader be guessed at by some passages of
Carneades (I say some passages, because I make not all
that he says, especially in the heat of disputation, mine),
partly in this discourse, and partly in some other [1] dialogues
betwixt the same speakers (though they treat not immediately of the elements) which have long lain by me,
and expect the entertainement that these present discourses will meet with. And indeed they will much
mistake me, that shall conclude from what I now publish,
that I am at defiance with chymistry, or would make my
readers so. I hope the *Specimina* I have lately published
of an attempt to shew the usefulness of chymical experiments to contemplative philosophers, will give those that
read them other thoughts of me, and I had a design (but
wanted opportunity) to publish with these papers an essay
I have lying by me, the greater part of which is apologetical
for one sort of chymists. And at least, as for those that
know me, I hope the pain I have taken in the fire will both
convince them that I am far from being an enemy to the
chymist's art (though I am no friend to many that disgrace
it by professing it), and persuade them to believe me when
I declare that I distinguish betwixt those chymists that
are either cheats, or but laborants, and the true adepti;
by whome could I enjoy their conversation, I would both
willingly and thankfully be instructed; especially concerning the nature and generation of metals: and possibly,
those that know how little I have remitted of my former
addictedness to make chymical experiments, will easily
believe that one of the chief designes of this sceptical discourse was, not so much to discredit chymistry, as to give

[1] The Dialogues here meant are those about Heat, Fire, Flame,
etc. (seen by two secretaries of the Royal Society), that the author
somewhere complaines to have been missing with other things of
his presently after the hasty removal of his goods by night in the
great fire of London.

an occasion and a kind of necessity to the more knowing
artists to lay aside a little of their over-great reservedness,
and either explicate or prove the chymical theory better
than ordinary chymists have done, or by enriching us
with some of their nobler secrets to evince that their art is
able to make amends even for the deficiencies of their
theory: and thus much I shall make bold to add, that
we shall much undervalue chymistry, if we imagine that
it cannot teach us things far more useful, not only to
physic but to philosophy, than those that are hitherto
know to vulgar chymists. And yet as for inferior spagyrists themselves, they have by their labours deserved so
well of the commonwealth of learning, that methinks 'tis
pity they should ever misse the truth which they have
so industriously sought. And though I be no admirer of
the theorical part of their art, yet my conjectures will
much deceive me, if the practical part be not hereafter
much more cultivated than hitherto it has been, and do
not both employ philosophy and philosophers, and hope to
make men such. Nor would I, that have been diverted
by other studies as well as affairs, be thought to pretend
being a profound spagyrist, by finding so many faults in
the doctrine wherein the generality of chymists scruples
not to acquiesce: for besides that 'tis most commonly far
easier to frame objections against any proposed hypothesis
than to propose an hypothesis not liable to objections,
(besides this I say) 'tis no such great matter, if whereas
beginners in chymistry are commonly at once imbued
with the theory and operations of their profession, I who
had the good fortune to learn the operations from illiterate
persons, upon whose credit I was not tempted to take up
any opinion about them, should consider things with lesse
prejudice, and consequently with other eyes than the
generality of learners; and should be more disposed to
accommodate the phænomena that occurred to me to other
notions than to those of the spagirists. And having at
first entertained a suspicion that the vulgar principles were
lesse general and comprehensive, or lesse considerately
deduced from chymical operations, than was believed, it
was not uneasie for me both to take notice of divers phæno-

mena, overlooked by prepossest persons, that seemed
not to suite so well with the hermetical doctrine; and to
devise some experiments likely to furnish me with objections against it, not known to many, that having practised
chymistry longer perchance than I have yet lived, may
have far more experience than I of particular processes.

To conclude, whether the notions I have proposed, and
the experiments I have communicated, be considerable,
or not, I willingly leave others to judge; and this only I
shall say for myself, that I have endeavoured to deliver
matters of fact so faithfully, that I may as well assist
the lesse skilful readers to examine the chymical hypothesis, as provoke the spagirical philosophers to illustrate
it: which if they do, and that either the chymical opinion,
or the peripatetic, or any other theory of the elements
differing from that I am most inclined to, shall be intelligibly explicated, and duly proved to me; what I have
hitherto discoursed will not hinder it from making a
proselyte of a person that loves fluctuation of judgment
little enough to be willing to be eased of it by anything
but error.

PHYSIOLOGICAL CONSIDERATIONS

TOUCHING THE EXPERIMENTS WONT TO BE
EMPLOYED TO EVINCE EITHER THE FOUR
PERIPATETICK ELEMENTS, OR THE THREE
CHYMICAL PRINCIPLES OF MIXT BODIES

PART OF THE FIRST DIALOGUE

I PERCEIVE that divers of my friends have thought it very
strange to hear me speak so irresolvedly, as I have been
wont to do, concerning those things which some take to
be the elements, and others to be the principles of all
mixt bodies. But I blush not to acknowledge that I
much less scruple to confess that I doubt when I do so,
than to profess that I know what I do not: and I should
have much stronger expectations than I dare yet entertain,
to see philosophy solidly established, if men would more
carefully distinguish those things that they know from
those that they ignore or do but think, and then explicate
clearly the things they conceive they understand, acknowledge ingenuously what it is they ignore, and profess so
candidly their doubts, that the industry of intelligent
persons might be set on work to make further enquiries,
and the easiness of less discerning men might not be
imposed on. But because a more particular accompt
will probably be expected of my unsatisfiedness not only
with the peripatetic, but with the chymical doctrine of
the primitive ingredients of bodies: it may possibly serve
to satisfy others of the excusableness of my dissatisfaction
to peruse the ensuing relation of what passed a while since
at a meeting of persons of several opinions, in a place that
need not here be named; where the subject, whereof
we have been speaking, was amply and variously discoursed of.

It was on one of the fairest dayes of this summer that the inquisitive Eleutherius came to invite me to make a visit with him to his friend Carneades. I readily consented to this motion, telling him that if he would but permit me to go first and make an excuse at a place not far off, where I had at that hour appointed to meet, but not about a business either of moment, or that could not well admit of a delay, I would presently wait on him, because of my knowing Carneades to be so conversant with nature and with furnaces, and so unconfined to vulgar opinions, that he would probably by some ingenious paradox or other give our mindes at least a pleasing exercise, and perhaps enrich them with some solid instruction. Eleutherius then first going with me to the place where my apology was to be made, I accompanied him to the lodging of Carneades, where when we were come, we were told by the servants that he was retired with a couple of friends (whose names they also told us) to one of the arbours in his garden, to enjoy under its coole shades a delightful protection from the yet troublesome heat of the sun.

Eleutherius being perfectly acquainted with that garden immediately led me to the arbour, and relying on the intimate familiarity that had been long cherished betwixt him and Carneades; in spite of my reluctancy to what might look like an intrusion upon his privacy, drawing me by the hand, he abruptly entered the arbour, where we found Carneades, Philoponus, and Themistius, sitting close about a little round table, on which, besides paper, pen, and inke, there lay two or three open books; Carneades appeared not at all troubled at this surprise, but rising from the table, received his friend with open looks and armes, and welcoming me also with his wonted freedom and civility, invited us to rest ourselves by him, which, as soon as we had exchanged with his two friends (who were ours also) the civilities accustomed on such occasions, we did. And he presently after we had seated ourselves, shutting the books that lay open, and turning to us with a smiling countenance, seemed ready to begin some such unconcerning discourse as is wont to pass, or rather waste, the time in promiscuous companies.

But Eleutherius guessing at what he meant to do, prevented him by telling him, I perceive, Carneades, by the books that you have been now shutting, and much more by the posture wherein I found persons so qualified to discourse of serious matters, and so accustomed to do it, that you three were, before our coming, engaged in some philosophical conference, which I hope you will either prosecute, and allow us to be partakers of, in recompense of the freedome we have used in presuming to surprise you, or else give us leave to repair the injury we should otherwise do you, by leaving you to the freedom we have interrupted, and punishing ourselves for our boldness by depriving ourselves of the happiness of your company. With these last words he and I rose up, as if we meant to be gone: but Carneades suddenly laying hold on his arme, and stopping him by it, smilingly told him, We are not so forward to lose good company as you seem to imagine; especially since you are pleased to desire to be present at what we shall say about such a subject as that you found us considering. For that, being the number of the elements, principles, or material ingredients of bodies, is an enquiry whose truth is of that importance, and of that difficulty, that it may as well deserve, as require, to be searched into by such skilful indagators of nature as yourselves. And therefore we sent to invite the bold and acute Leucippus to lend us some light by his atomical paradox, upon which we expected such pregnant hints, that 'twas not without a great deal of trouble that we had lately word brought us that he was not to be found; and we had likewise begged the assistance of your presence and thoughts, had not the messenger we employed to Leucippus informed us that as he was going he saw you both pass by towards another part of the town; and this frustrated expectation of Leucippus his company, who told me but last night that he would be ready to give me a meeting where I pleased to-day, having very long suspended our conference about the freshly mentioned subject, it was so newly begun when you came in, that we shall scarce need to repeat anything to acquaint you with what had passed betwixt us before your arrival, so that I cannot but look upon it as a fortunate accident that you should come so seasonably, to be not hearers alone, but we hope interlocutors at our conference. For we shall not only allow of your presence at it, but desire your assistance in it; which I add both for other reasons, and because though these learned gentlemen (says he, turning to his two friends) need not fear to discourse before any auditory, provided it be intelligent enough to understand them, yet for my part (continues he with a new smile) I shall not dare to vent my unpremeditated thoughts before two such critics, unless by promising to take your turnes of speaking, you will allow me mine of quarrelling with what has been said. He and his friends added divers things to convince us that they were both desirous that we should hear them, and resolved against our doing so, unless we allowed them sometimes to hear us. Eleutherius, after having a while fruitlessly endeavoured to obtain leave to be silent, promised he would not be so alwayes, provided that he were permitted according to the freedom of his genius and principles to side with one of them in the managing of one argument, and, if he saw cause, with his antagonist, in the prosecution of another, without being confined to stick to any one party or opinion, which was after some debate accorded him. But, I conscious of my own disabilities, told them resolutely that I was as much more willing, as more fit, to be a hearer than a speaker among such knowing persons, and on so abstruse a subject. And that therefore I beseeched them without necessitating me to proclaim my weaknesses, to allow me to lessen them by being a silent auditor of their discourses: to suffer me to be at which I could present them no motive, save that their instructions would make them in me a more intelligent admirer. I added that I desired not to be idle whilst they were imployed, but would if they pleased, by writing down in shorthand what should be delivered, preserve discourses that I knew would merit to be lasting. At first Carneades and his two friends utterly rejected this motion; and all that my resoluteness to make use of my ears, not tongue, at their debates could do, was to make them acquiesce in the proposition of Eleutherius, who thinking himself concerned, because he brought me thither, to afford me some faint assistance, was content that I should register their arguments, that I might be the better able after the conclusion of their conference to give them my sense upon the subject of it (the number of elements or principles), which he promised I should do at the end of the present debates, if time would permit, or else at our next meeting. And this being by him undertaken in my name, though without my consent, the company would by no means receive my protestation against it, but casting, all at once, their eyes on Carneades, they did by that and their unanimous silence, invite him to begin; which (after a short pause, during which he turned himself to Eleutherius and me) he did in this manner.

Notwithstanding the subtile reasonings I have met with in the books of the peripatetics, and the pretty experiments that have been shewed me in the laboratories of chymists, I am of so diffident or dull a nature, as to think that if neither of them can bring more cogent arguments to evince the truth of their assertion than are wont to be brought, a man may rationally enough retain some doubts concerning the very number of those material ingredients of mixt bodies, which some would have us call elements, and others principles. Indeed when I considered that the tenets concerning the elements are as considerable amongst the doctrines of natural philosophy, as the elements themselves are among the bodies of the universe, I expected to find those opinions solidly established, upon which so many others are superstructed. But when I took the pains impartially to examine the bodies themselves that are said to result from the blended elements, and to torture them into a confession of their constituent principles, I was quickly induced to think that the number of the elements has been contended about by philosophers with more earnestness than success. This unsatisfiedness of mine has been much wondered at by these two gentlemen (at which words he pointed at Themistius and Philoponus), who though they differ almost as much betwixt themselves about the question we are to consider, as I do from either of them, yet they both agree very well in this, that there is a determinate number of such ingredients as I was just now speaking of, and that what that number is I say not, may be (for what may not such as they persuade?), but is wont to be clearly enough demonstrated both by reason and experience. This has occasioned our present conference. For our discourse this afternoon, having fallen from one subject to another, and at length settled on this, they proffered to demonstrate to me, each of them the truth of his opinion, out of both the topics that I have freshly named. But on the former (that of reason strictly so taken) we declined insisting at the present, lest we should not have time enough before supper to go through the reasons and experiments too. The latter of which we unanimously thought the most requisite to be seriously examined. I must desire you then to take notice, gentlemen (continued Carneades), that my present business doth not oblige me so to declare my own opinion on the subject in question as to assert or deny the truth either of the peripatetic or the chymical doctrine concerning the number of the elements, but only to shew you that neither of these doctrines hath been satisfactorily proved by the arguments commonly alledged on its behalfe. So that if I really discern (as perhaps I think I do) that there may be a more rational account than ordinary, given of one of these opinions, I am left free to declare myself of it, notwithstanding my present engagement, it being obvious to all your observation, that a solid truth may be generally maintained by no other than incompetent arguments. And to this declaration I hope it will be needless to add, that my task obliges me not to answer the arguments that may be drawn either for Themistius's or Philoponus's opinion from the topic of reason, as opposed to experiments; since 'tis these only that I am to examine, and not all these neither, but such of them alone as either of them shall think fit to insist on, and as have hitherto been wont to be brought either to prove that 'tis the four peripatetic elements, or that 'tis the three chymical principles that all compounded bodies consist of. These things (adds Carneades) I thought myself obliged to

premise, partly lest you should do these gentlemen (pointing at Themistius and Philoponus, and smiling on them)
the injury of measuring their parts by the arguments they
are ready to propose, the lawes of our conference confining
them to make use of those that the vulgar of philosophers (for even of them there is a vulgar) has drawn up
to their hands ; and partly that you should not condemn
me of presumption for disputing against persons over
whom I can hope for no advantage, that I must not derive
from the nature or rules of our controversy, wherein I
have but a negative to defend, and wherein too I am like
on several occasions to have the assistance of one of my
disagreeing adversaries against the other.

Philoponus and Themistius soon returned this com-
pliment with civilities of the like nature, in which Eleutherius perceiving them engaged, to prevent the further
loss of that time of which they were not like to have very
much to spare, he minded them that their present business was not to exchange compliments, but arguments:
and then addressing his speech to Carneades, I esteem it
no small happiness (says he) that I am come here so
luckily this evening. For I have been long disquieted
with doubts concerning this very subject which you are
now ready to debate. And since a question of this importance is to be now discussed by persons that maintain
such variety of opinions concerning it, and are both so
able to enquire after truth, and so ready to embrace
it by whomsoever and on what occasion soever it is
presented them; I cannot but promise myself that I
shall before we part, either lose my doubts or the hopes of
ever finding them resolved; Eleutherius paused not here;
but to prevent their answer, added almost in the same
breath; and I am not a little pleased to find that you are
resolved on this occasion to insist rather on experiments
than syllogismes. For I, and no doubt you, have long
observed, that those dialectical subtleties, that the schoolmen too often employ about physiological mysteries, are
wont much more to declare the wit of him that uses them,
than increase the knowledge or remove the doubts of
sober lovers of truth. And such captious subtleties do

indeed often puzzle and sometimes silence men, but rarely satisfy them. Being like the tricks of jugglers, whereby men doubt not but they are cheated, though oftentimes they cannot declare by what flights they are imposed on. And therefore I think you have done very wisely to make it your business to consider the phænomena relating to the present question, which have been afforded by experiments, especially since it might seem injurious to our senses, by whose mediation we acquire so much of the knowledge we have of things corporal, to have recourse to far-fetched and abstracted ratiocinations, to know what are the sensible ingredients of those sensible things that we daily see and handle, and are supposed to have the liberty to untwist (if I may so speak) into the primitive bodies they consist of. He annexed that he wished therefore they would no longer delay his expected satisfaction, if they had not, as he feared they had, forgotten something preparatory to their debate; and that was to lay down what should be all along understood by the word principle or element. Carneades thanked him for his admonition, but told him that they had not been unmindful of so requisite a thing. But that being gentlemen and very far from the litigious humour of loving to wrangle about words, or terms, or notions as empty, they had before his coming in readily agreed promiscuously to use when they pleaded, elements and principles as terms equivalent: and to understand both by the one and the other, those primitive and simple bodies of which the mixt ones are said to be composed, and into which they are ultimately resolved. And upon the same account (he added) we agreed to discourse of the opinions to be debated, as we have found them maintained by the generality of the assertors of the four elements of the one party, and of those that receive the three principles on the other, without tying ourselves to enquire scrupulously what notion either Aristotle or Paracelsus, or this or that interpreter or follower of either of those great persons, framed of elements or principles; our design being to examine, not what these or those writers thought or taught, but what we find to be the obvious and most general opinion of those who are willing to be accounted favourers of the peripatetic or chymical doctrine concerning this subject.

I see not (says Eleutherius) why you might not immediately begin to argue, if you were but agreed which of your two friendly adversaries shall be first heard. And it being quickly resolved on that Themistius should first propose the proofs for his opinion, because it was the antienter, and the more general, he made not the company expect long before he thus addressed himself to Eleutherius, as to the person least interested in the dispute.

If you have taken sufficient notice of the late confession which was made by Carneades, and which (though his civility dressed it up in complimental expressions) was exacted of him by his justice, I suppose you will be easily made sensible, that I engage in this controversie with great and peculiar disadvantages, besides those which his parts and my personal disabilities would bring to any other cause to be maintained by me against him. For he justly apprehending the force of truth, though speaking by no better a tongue than mine, has made it the chief condition of our duel, that I should lay aside the best weapons I have, and those I can best handle; whereas if I were allowed the freedom, in pleading for the four elements, to employ the arguments suggested to me by reason to demonstrate them, I should almost as little doubt of making you a proselyte to those unsevered teachers, Truth and Aristotle, as I do of your candour and your judgment. And I hope you will however consider, that that great favourite and interpreter of nature, Aristotle, who was (as his *Organum* witnesses) the greatest master of logic that ever lived, disclaimed the course taken by other petty philosophers (antient and modern), who not attending the coherence and consequences of their opinions, are more solicitous to make each particular opinion plausible independently upon the rest, than to frame them all so, as not only to be consistent together, but to support each other. For that great man in his vast and comprehensive intellect, so framed each of his notions, that being curiously adapted into one systeme, they need not each of them any other defence than that which their mutual coherence gives them: as 'tis in an arch, where each single stone, which if severed from the rest would be perhaps defenceless, is sufficiently secured by the solidity and entireness of the whole fabric of which it is a part. How justly this may be applied to the present case, I could easily shew you, if I were permitted to declare to you, how harmonious Aristotle's doctrine of the elements is with his other principles of philosophy; and how rationally he has deduced their number from that of the combinations of the four first qualities from the kinds of simple motion belonging to simple bodies, and from I know not how many other principles and phænomena of nature, which so conspire with his doctrine of the elements, that they mutually strengthen and support each other. But since 'tis forbidden me to insist on reflections of this kind, I must proceed to tell you, that though the assertors of the four elements value reason so highly, and are furnished with arguments enough drawn from thence, to be satisfied that there must be four elements, though no man had ever yet made any sensible trial to discover their number, yet they are not destitute of experience to satisfie others that are wont to be more swayed by their senses than their reason. And I shall proceed to consider the testimony of experience, when I shall have first advertised you, that if men were as perfectly rational as 'tis to be wished they were, this sensible way of probation would be as needless as 'tis wont to be imperfect. For it is much more high and philosophical to discover things *a priore* than *a posteriore*. And therefore the peripatetics have not been very solicitous to gather experiments to prove their doctrines, contenting themselves with a few only, to satisfy those that are not capable of a nobler conviction. And indeed they employ experiments rather to illustrate than to demonstrate their doctrines, as astronomers use sphæres of pasteboard, to descend to the capacities of such as must be taught by their senses, for want of being arrived to a clear apprehension of purely mathematical notions and truths. I

speak thus, Eleutherius (adds Themistius), only to do right
to reason, and not out of diffidence of the experimental
proof I am to alledge. For though I shall name but one,
yet it is such a one as will make all other appear as needless as itself will be found satisfactory. For if you but
consider a piece of green wood burning in a chimney, you
will readily discern in the disbanded parts of it the four
elements, of which we teach it and other mixt bodies to be
composed. The fire discovers itself in the flame by its own
light; the smoake by ascending to the top of the chimney,
and there readily vanishing into air, like a river losing
itself in the sea, sufficiently manifests to what element it
belongs and gladly returnes. The water in its own form
boiling and hissing at the ends of the burning wood
betrays itself to more than one of our senses; and the
ashes by their weight, their firiness, and their dryness,
put it past doubt that they belong to the element of
earth. If I spoke (continues Themistius) to less knowing
persons, I would perhaps make some excuse for building
upon such an obvious and easie analysis, but 'twould be, I
fear, injurious, not to think such an apology needless to
you, who are too judicious either to think it necessary that
experiments to prove obvious truths should be far-fetched,
or to wonder that among so many mixt bodies that are
compounded of the four elements, some of them should
upon a slight analysis manifestly exhibite the ingredients
they consist of. Especially since it is very agreeable to the
goodness of nature to disclose, even in some of the most
obvious experiments that men make, a truth so important and so requisite to be taken notice of by them.
Besides that our analysis by how much the more obvious
we make it, by so much the more suitable it will be to the
nature of that doctrine which 'tis alledged to prove, which
being as clear and intelligible to the understanding as
obvious to the sense, 'tis no marvel the learned part of
mankind should so long and so generally imbrace it. For
this doctrine is very different from the whimseys of
chymists and other modern innovators, of whose hypotheses we may observe, as naturalists do of less perfect
animals, that as they are hastily formed, so they are

commonly short-lived. For so these, as they are often
framed in one week, are perhaps thought fit to be laughed
at the next; and being built perchance but upon two or
three experiments are destroyed by a third or fourth,
whereas the doctrine of the four elements was framed by
Aristotle after he had leasurely considered those theories
of former philosophers which are now with great
applause revived as discovered by these latter ages; and
had so judiciously detected and supplied the errors and
defects of former hypotheses concerning the elements,
that his doctrine of them has been ever since deservedly
embraced by the lettered part of mankind: all the philosophers that preceded him having in their several ages
contributed to the compleatness of this doctrine, as those
of succeeding times have acquiesced in it. Nor has an
hypothesis, so deliberately and maturely established, been
called in question till in the last century Paracelsus and
some few other sooty empirics, rather than (as they are
fain to call themselves) philosophers, having their eyes
darkened, and their braines troubled with the smoak of
their own furnaces, began to rail at the peripatetic
doctrine, which they were too illiterate to understand,
and to tell the credulous world, that they could see but
three ingredients in mixt bodies; which to gain themselves the repute of inventors, they endeavoured to disguise by calling them, instead of earth, and fire, and
vapour, salt, sulphur, and mercury; to which they gave
the canting title of hypostatical principles. But when they
came to describe them, they shewed how little they understood what they meant by them, by disagreeing as much
from one another, as from the truth they agreed in oppos-
ing: for they deliver their hypotheses as darkly as their
processes; and 'tis almost as impossible for any sober man
to find their meaning, as 'tis for them to find their elixir.
And indeed nothing has spread their philosophy, but their
great brags and undertakings; notwithstanding all which
(says Themistius smiling), I scarce know anything they
have performed worth wondering at, save that they have
been able to draw Philoponus to their party, and to engage
him to the defence of an unintelligible hypothesis, who

knowes so well as he does, that principles ought to be like diamonds, as well very clear as perfectly solid.

Themistius having after these last words declared by his silence that he had finished his discourse, Carneades addressing himself, as his adversary had done, to Eleutherius, returned this answer to it. I hoped for a demonstration, but I perceive Themistius hopes to put me off with an harangue, wherein he cannot have given me a greater opinion of his parts, than he has given me distrust for his hypothesis, since for it even a man of such learning can bring no better arguments. The rhetorical part of his discourse, though it make not the least part of it, I shall say nothing to, designing to examine only the argumentative part, and leaving it to Philoponus to answer those passages wherein either Paracelsus or chymists are concerned: I shall observe to you, that in what he has said besides, he makes it his business to do these two things. The one to propose and make out an experiment to demonstrate the common opinion about the four elements; and the other, to insinuate divers things which he thinks may repair the weakness of his argument, from experience, and upon other accounts bring some credit to the otherwise defenceless doctrine he maintains.

To begin then with his experiment of the burning wood, it seems to me to be obnoxious to not a few considerable exceptions.

And first, if I would now deal rigidly with my adversary, I might here make a great question of the very way of probation which he and others employ, without the least scruple, to evince that the bodies commonly called mixt are made up of earth, air, water, and fire, which they are pleased also to call elements; namely that upon the supposed analysis made by the fire, of the former sort of concretes, there are wont to emerge bodies resembling those which they take for the elements. For not to anticipate here what I foresee I shall have occasion to insist on, when I come to discourse with Philoponus concerning the right that fire has to pass for the proper and universal instrument of analysing mixt bodies, not to anticipate that, I say, if I were disposed to wrangle, I might alledge, that by Themistius his experiment it would appear rather that those he calls elements are made of those he calls mixt bodies, than mixt bodies of the elements. For in Themistius's analysed wood, and in other bodies dissipated and altered by the fire, it appears, and he confesses, that which he takes for elementary fire and water are made out of the concrete; but it appears not that the concrete was made up of fire and water. Nor has either he, or any man, for ought I know, of his persuasion, yet proved that nothing can be obtained from a body by the fire that was not pre-existent in it.

At this unexpected objection, not only Themistius, but the rest of the company appeared not a little surprised; but after a while Philoponus conceiving his opinion, as well as that of Aristotle, concerned in that objection, You cannot sure (says he to Carneades) propose this difficulty, not to call it cavil, otherwise than as an exercise of wit, and not as laying any weight upon it. For how can that be separated from a thing that was not existent in it? When, for instance, a refiner mingles gold and lead, and exposing this mixture upon a cuppel to the violence of the fire, thereby separates it into pure and refulgent gold and lead (which driven off together with the dross of the gold is thence called *lythargyrium auri*), can any man doubt that sees these two so differing substances separated from the mass, that they were existent in it before it was committed to the fire?

I should (replies Carneades) allow your argument to prove something, if, as men see the refiners commonly take beforehand both lead and gold to make the mass you speak of, so we did see nature pull down a parcel of the element of fire, that is fancied to be placed I know not how many thousand leagues off, contiguous to the orb of the moon, and to blend it with a quantity of each of the three other elements, to compose every mixt body, upon whose resolution the fire presents us with fire, and earth, and the rest. And let me add, Philoponus, that to make your reasoning cogent, it must be first proved, that the fire does only take the elementary ingredients asunder, without otherwise altering them. For else 'tis obvious, that bodies may afford substances which were not pre-existent in them; as flesh too long kept produces maggots, and old cheese mites, which I suppose you will not affirm to be ingredients of those bodies. Now that fire does not alwayes barely separate the elementary parts, but some-times at least alter also the ingredients of bodies, if I did not expect ere long a better occasion to prove it, I might make probable out of your very instance, wherein there is nothing elementary separated by the great violence of the refiner's fire: the gold and lead which are the two ingredients separated upon the analysis being con-fessedly yet perfectly mixt bodies, and the litharge being lead indeed, but such lead as is differing in consist-ence and other qualities from what it was before. To which I must add that I have sometimes seen, and so questionless have you much oftener, some parcels of glasse adhering to the test or cuppel, and this glass, though emergent as well as the gold or litharge upon your analysis, you will not I hope allow to have been a third ingredient of the mass out of which the fire produced it.

Both Philoponus and Themistius were about to reply, when Eleutherius apprehending that the prosecution of this dispute would take up time which might be better employed, thought fit to prevent them by saying to Carneades: You made at least half a promise, when you first proposed this objection, that you would not (now at least) insist on it, nor indeed does it seem to be of absolute necessity to your cause that you should. For though you should grant that there are elements, it would not follow that there must be precisely four. And therefore I hope you will proceed to acquaint us with your other and more considerable objections against Themistius's opinion, especially since there is so great a disproportion in bulke betwixt the earth, water, and air, on the one part, and those little parcels of resembling substances that the fire separates from concretes on the other part, that I can scarce think that you are serious, when to lose no advantage against your adversary, you seem to deny it to be rational to conclude these great simple bodies to be the elements, and not the products of compounded ones.

What you alledge (replies Carneades) of the vastness of the earth and water, has long since made me willing to allow them to be the greatest and chief masses of matter to be met wi h here below: but I think I could shew you, if you would give me leave, that this will prove only that the elements, as you call them, are the chief bodies that make up the neighbouring part of the world, but not that they are such ingredients as every mixt body must consist of. But since you challenge me of something of a promise, though it be not an entire one, yet I shall willingly performe it. And indeed I intended not, when I first mentioned this objection, to insist on it at present against Themistius (as I plainly intimated in my way of proposing it), being only desirous to let you see, that though I discerned my advantages, yet I was willing to forego some of them rather than appear a rigid adversary of a cause so weak, that it may with safety be favourably dealt with. But I must here profess, and desire you to take notice of it, that though I pass on to another argument, it is not because I think this first invalid. For you will find in the progress of our dispute, that I had some reason to question the very way of probation imployed both by peripatetics and chymists, to evince the being and number of the elements. For that there are such, and that they are wont to be separated by the analysis made by fire, is indeed taken for granted by both parties, but has not (for ought I know) been so much as plausibly attempted to be proved by either. Hoping then that when we come to that part of our debate, wherein considerations relating to this matter are to be treated of, you will remember what I have now said, and that I do rather for a while suppose than absolutely grant the truth of what I have questioned, I will proceed to another objection.

And hereupon Eleutherius having promised him not to be unmindful, when time should serve, of what he had declared.

I consider then (says Carneades), in the next place, that there are divers bodies out of which Themistius will not prove in haste that there can be so many elements as four extracted by the fire. And I should perchance trouble him if I should ask him what peripatetic can shew us (I say not, all the four elements, for that would be too rigid a question, but) any one of them extracted out of gold by any degree of fire whatsoever. Nor is gold the only bodie in nature that would puzzle an Aristotelian, (that is no more) to analyse by the fire into elementary bodies, since, for ought I have yet observed, both silver and calcined Venetian talc, and some other concretes, not necessary here to be named, are so fixed, that to reduce any of them into four heterogeneous substances has hitherto proved a task much too hard, not only for the disciples of Aristotle, but those of Vulcan, at least, whilst the latter have employed only fire to make the analysis.

The next argument (continues Carneades) that I shall urge against Themistius's opinion shall be this, That as there are divers bodies whose analysis by fire cannot reduce them into so many heterogeneous substances or ingredients as four, so there are others which may be reduced into more, as the blood (and divers other parts) of men and other animals, which yield when analysed five distinct substances, phlegme, spirit, oile, salt, and earth, as experience has shewn us in distilling man's blood, harts-horns, and divers other bodies that belonging to the animal-kingdom abound with not uneasily sequestrable salt.

THE SCEPTICAL CHYMIST

THE FIRST PART

I AM (says Carneades) so unwilling to deny Eleutherius anything, that though before the rest of the company I am resolved to make good the part I have undertaken of a sceptic, yet I shall readily, since you will have it so. lay aside for a while the person of an adversary to the peripatetics and chymists; and before I acquaint you with my objections against their opinions, acknowledge to you what may be (whether truly or not) tolerably enough added, in favour of a certain number of principles of mixt bodies, to that grand and known argument from the analysis of compound bodies, which I may possibly hereafter be able to confute.

And that you may the more easily examine and the better judge of what I have to say, I shall cast it into a pretty number of distinct propositions, to which I shall not premise anything; because I take it for granted, that you need not be advertised that much of what I am to deliver, whether for or against a determinate number of ingredients of mixt bodies, may be indifferently applied to the four peripatetic elements, and the three chymical principles, though divers of my objections will more peculiarly belong to these last named, because the chymical hypothesis seeming to be much more countenanced by experience than the other, it will be expedient to insist chiefly upon the disproving of that; especially since most of the arguments that are imployed against it, may, by a little variation, be made to conclude, at least as strongly, against the less plausible, Aristotelian doctrine.

To proceed then to my propositions I shall begin with this, that—

PROPOSITION I.—*It seems not absurd to conceive that at the first production of mixt bodies, the universal matter whereof they among other parts of the universe consisted, was actually divided into little particles of several sizes and shapes variously moved.*

This (says Carneades) I suppose you will easily enough allow. For besides that which happens in the generation, corruption, nutrition, and wasting of bodies, that which we discover partly by our microscopes of the extream littleness of even the scarce sensible parts of concretes, and partly by the chymical resolutions of mixt bodies, and by divers other operations of spagirical fires upon them, seems sufficiently to manifest their consisting of parts very minute and of differing figures. And that there does also intervene a various local motion of such small bodies, will scarce be denied; whether we chuse to grant the origine or concretions assigned by Epicurus, or that related by Moses. For the first, as you well know, supposes not only all mixt bodies, but all others, to be produced by the various and casual occursions of atomes, moving themselves to and fro by an internal principle in the immense or rather infinite vacuum. And as for the inspired historian, he, informing us that the great and wise Author of things did not immediately create plants, beasts, birds, etc., but produced them out of those portions of the pre-existent, though created, matter, that he calls water and earth, allows us to conceive that the constituent particles whereof these new concretes were to consist, were variously moved in order to their being connected into the bodies they were, by their various coalitions and textures, to compose.

But (continues Carneades) presuming that the first proposition needs not be longer insisted on, I will pass on to the second, and tell you that—

PROPOSITION II.—*Neither is it impossible that of these minute particles divers of the smallest and neighbouring ones were here and there associated into minute masses or clusters, and did by their coalitions constitute great store of such little primary concretions or masses as were not easily dissipable into such particles as composed them.*

To what may be deduced, in favour of this assertion from the nature of the thing itself, I will add something out of experience, which though I have not known it used to such a purpose, seems to me more fairly to make out that there may be elementary bodies, than the more questionable experiments of peripatetics and chymists prove that there are such. I consider then that gold will mix and be colliquated not only with silver, copper, tin and lead, but with antimony, *regulus martis* and many other minerals, with which it will compose bodies very differing both from gold, and the other ingredients of the resulting concretes. And the same gold will also by common *aqua regis*, and (I speak it knowingly) by divers other menstruums, be reduced into a seeming liquor, insomuch that the corpuscles of gold will, with those of the menstruum, pass through cap-paper, and with them also coagulate into a crystalline salt. And I have further tried, that with a small quantity of a certain saline substance I prepared, I can easily enough sublime gold into the form of red crystals of a considerable length; and many other wayes may gold be disguised, and help to constitute bodies of very differing natures both from it and from one another, and nevertheless be afterward reduced to the self-same numerical, yellow, fixt, ponderous, and malleable gold it was before its commixture. Nor is it only the fixedst of metals, but the most fugitive, that I may employ in favour of our proposition: for quicksilver will with divers metals compose an amalgam, with divers menstruums it seems to be turned into a liquor, with *aqua fortis* it will be brought into either a red or white powder or precipitate, with oil of vitriol into a pale yellow one, with sulphur it will compose a blood-red and volatile cinaber, with some saline bodies it will ascend in form of a salt which will be dissoluble in water; with regulus of antimony and silver I have seen it sublimed into a kinde of crystals, with another mixture I reduced it into a malleable body, into a hard and brittle substance by another: and some there are who affirm, that by proper additaments they can reduce quicksilver into oil, nay into glass, to mention no more. And yet out of all these exotic compounds, we may recover the very same running mercury that was the main ingredient of them, and was so disguised in them. Now the reason (proceeds Carneades) that I have represented these things concerning gold and quicksilver, is, that it may not appear absurd to conceive, that such little primary masses or clusters as our proposition mentions, may remain undissipated, notwithstanding their entering into the composition of various concretions, since the corpuscle of gold and mercury, though they be not primary concretions of the most minute particles of matter, but confessedly mixt bodies, are able to concure plentifully to the composition of several very differing bodies, without losing their own nature or texture, or having their cohesion violated by the divorce of their associated parts or ingredients.

Give me leave to add (says Eleutherius) on this occasion, to what you now observed, that as confidently as some chymists, and other modern innovators in philosophy are wont to object against the peripatetics, that from the mixture of their four elements there could arise but an inconsiderable variety of compound Bodies; yet if the Aristotelians were but half as well versed in the works of nature as they are in the writings of their master, the proposed objection would not so calmly triumph, as for want of experiments they are fain to suffer it to do. For if we assigne to the corpuscles, whereof each element consists, a peculiar size and shape, it may easily enough be manifested, that such differingly figured corpuscles may be mingled in such various proportions, and may be connected so many several ways, that an almost incredible number of variously qualified concretes may be composed of them. Especially since the corpuscles of one element may barely, by being associated among themselves, make up little masses of differing size and figure from their constituent parts; and since also to the strict union of such minute bodies there seems oftentimes nothing requisite, besides the bare contact of a great part of their surfaces. And how great a variety of phænomena the same matter, without the addition of any other, and only several ways disposed or contexed, is able to exhibit, may partly appear by the multitude of differing engins which by the contrivances of skilful mechanilians, and the dexterity of expert workmen, may be made of iron alone. But in our present case being allowed to deduce compound bodies from four very differently qualified sorts of matter, he who shall but consider what you freshly took notice of concerning the new concretes resulting from the mixture of incorporated minerals, will scarce doubt but that the four elements managed by nature's skill may afford a multitude of differing compounds.

I am thus far of your minde (says Carneades) that the Aristotelians might with probability deduce a much greater number of compound bodies from the mixture of their four elements, than according to their present hypothesis they can, if instead of vainly attempting to deduce the variety and proprieties of all mixt bodies from the combinations and temperaments of the four elements, as they are (among them) endowed with the four first qualities, they had endeavoured to do it by the bulk and figure of the smallest parts of those supposed elements. For from these more catholic and fruitful accidents of the elementary matter may spring a great variety of textures, upon whose account a multitude of compound bodies may very much differ from one another. And what I now observe touching the four peripatetic elements, may be also applied, *mutatis mutandis* (as they speak), to the chymical principles. But (to take notice of that by the by) both the one and the other must, I fear, call in to their assistance something that is not elementary, to excite or regulate the motion of the parts of the matter, and dispose them after the manner requisite to the constitution of particular concretes. For that otherwise they are like to give us but a very imperfect account of the origine of very many mixt bodies, it would, I think, be no hard matter to persuade you, if it would not spend time, and were no digression, to examine, what they are wont to alledge of the origine of the textures and qualities of mixt bodies from a certain substantial form, whose origination they leave more obscure than what it is assumed to explicate.

But to proceed to a new proposition.

PROPOSITION III.—*I shall not peremptorily deny, that from most of such mixt bodies as partake either of animal or vegetable nature, there may by the help of the fire be actually obtained a determinate number (whether three, four, or five, or fewer or more) of substances, worthy of differing denominations.*

Of the experiments that induce me to make this concession, I am like to have occasion enough to mention several in the prosecution of my discourse. And therefore, that I may not hereafter be obliged to trouble you and myself with needless repetitions, I shall now only desire you to take notice of such experiments when they shall be mentioned, and in your thoughts referre them hither.

To these three concessions I have but this fourth to add, that—

PROPOSITION IV.—*It may likewise be granted, that those distinct substances, which concretes generally either afford or are made up of, may without very much inconvenience be called the elements or principles of them.*

When I said, *without very much inconvenience,* I had in my thoughts that sober admonition of Galen, *Cum de re constat, de verbis non est litigandum.* And therefore also I scruple not to say *elements* or *principles,* partly because the chymists are wont to call the ingredients of mixt bodies, *principles,* as the Aristotelians name them *elements ;* I would here exclude neither. And, partly, because it seems doubtful whether the same ingredients may not be called *principles :* as not being compounded of any more primary bodies: and *elements,* in regard that all mixt bodies are compounded of them. But I thought it requisite to limit my concession by premising the words *very much* to the word *inconvenience,* because that though the inconvenience of calling the distinct substances, mentioned in the proposition *elements* or *principles,* be not very great, yet that it is impropriety of speech, and consequently in a matter of this moment not to be altogether overlooked, you will perhaps think, as well as I, by that time you shall have heard the following part of my discourse, by which you will best discern what construction to put upon the former propositions, and how far they may be looked upon as things that I concede as true, etc., how far as things I only represent as specious enough to be fit to be considered.

And now, Eleutherius (continues Carneades), I must resume the person of a sceptic, and as such, propose some part of what may be either disliked, or at least doubted of in the common hypothesis of the chymists; which if I examine with a little the more freedom, I hope I need not desire you (a person to whom I have the happiness of being so well known) to look upon it as something more suitable to the employment whereto the company has, for this meeting, doomed me, than either to my humour or my custom.

Now though I might present you many things against the vulgar chymical opinion of the three principles and the experiments wont to be alleged as demonstrations of it, yet those I shall at present offer you may be conveniently enough comprehended in four capital considerations; touching all which I shall only premise this in general, That since it is not my present task so much to assert an hypothesis of my own, as to give an account wherefore I suspect the truth of that of the chymists, it ought not to be expected that all my objections should be of the most cogent sort, since it is reason enough to doubt of a proposed opinion, that there appears no cogent reason for it.

To come then to the objections themselves; I consider in the first place, that notwithstanding what common chymists have proved or taught, it may reasonably enough be doubted, how far, and in what sense, fire ought to be esteemed the genuine and universal instrument of analysing mixt bodies.

This doubt, you may remember, was formerly mentioned, but so transiently discoursed of, that it will now be fit to insist upon it, and manifest that it was not so inconsiderately proposed as our adversaries then imagined.

But, before I enter any further into this disquisition, I cannot but here take notice, that it were to be wished our chymists had clearly informed us what kind of division of bodies by fire must determine the number of the elements: For it is nothing near so easy as many seem to think, to determine distinctly the effects of heat, as I could easily manifest, if I had leasure to shew you how much the operations of fire may be diversified by circumstances. But not wholly to pass by a matter of this importance, I will first take notice to you that guajacum (for instance) burnt with an open fire in a chimney, is sequestred into ashes and soot, whereas the same wood distilled in a retort does yield far other heterogeneities (to use the Helmontian expression), and is resolved into oil, spirit, vinegar, water and charcoal; the last of which to be reduced into ashes, requires the being farther calcined than it can be in a close vessel: besides having kindled amber, and held a clean silver spoon, or some other concave and smooth vessel, over the smoak of its flame, I observed the soot into which that fume condensed to be very differing from anything that I had observed to proceed from the steam of amber purposely (for that is not usual) distilled *per se* in close vessels. Thus having, for trial's sake, kindled camphire and catcht the smoak that copiously ascended out of the flame, it condensed into a black and unctuous soot, which would not have been guessed by the smell or other properties to have proceeded from camphire: whereas having (as I shall other, where more fully declare) exposed a quantity of that fugitive concrete to a gentle heat in a close glass vessel, it sublimed up without seeming to have lost anything of its whiteness, or its nature, both which it retained, though afterwards I so encreased the fire as to bring it to fusion. And,

besides camphire, there are divers other bodies (that
I elsewhere name) in which the heat in close vessels is not
wont to make any separation of heterogeneities, but only
a comminution of parts, those that rise first being
homogeneal with the others, though subdivided into
smaller particles: whence sublimations have been styled,
The pestles of the chymists. But not here to mention
what I elsewhere take notice of, concerning common
brimstone once or twice sublimed, that exposed to a
moderate fire in subliming-pots, it rises all into dry, and
almost tasteless, flowers; whereas being exposed to a
naked fire it affords store of a saline and fretting liquor:
not to mention this, I say, I will further observe to you,
that as it is considerable in the analysis of mixt bodies,
whether the fire act on them when they are exposed to the
open air, or shut up in close vessels, so is the degree of fire,
by which the analysis is attempted, of no small moment.
For a milde *balneum* will sever unfermented blood (for
instance) but into phlegme and *caput mortuum*, the latter
whereof (which I have sometimes had), hard, brittle, and of
divers colours (transparent almost like tortoise-shell),
pressed by a good fire in a retort yields a spirit, an oil or
two, and a volatile salt, besides another *caput mortuum*. It
may be also pertinent to our present designe, to take notice
of what happens in the making and distilling of soap; for by
one degree of fire the salt, the water, and the oil or grease,
whereof that factitious concrete is made up, being boiled
up together are easily brought to mingle and incorporate
into one mass; but by another and further degree of heat
the same mass may be again divided into an oleagenous
and aqueous, a saline, and an earthy part. And so we
may observe that impure silver and lead being exposed
together to a moderate fire will thereby be colliquated into
one mass, and mingle *per minima*, as they speak; whereas
a much vehementer fire will drive or carry off the baser
metals (I mean the lead, and the copper or other alloy)
from the silver, though not, for ought appears, separate
them from one another. Besides, when a vegetable
abounding in fixt salt is analysed by a naked fire, as one
degree of heat will reduce it into ashes (as the chymists

themselves teach us), so, by only a further degree of fire, those ashes may be vitrified and turned into glass.   I will not stay to examine how far a mere chymist might on this occasion demand, if it be lawful for an Aristotelian to make ashes (which he mistakes for mere earth) pass for an element, because by one degree of fire it may be produced, why a chymist may not upon the like principle argue that glass is one of the elements of many bodies, because that also may be obtained from them, barely by the fire?   I will not, I say, lose time to examine this, but observe that by a method of applying the fire, such similar bodies may be obtained from a concrete, as chymists have not been able to separate, either by barely burning it in an open fire, or by barely distilling it in close vessels.   For to me it seems very considerable, and I wonder that men have taken so little notice of it, that I have not by any of the common wayes of distillation in close vessels seen any separation made of such a volatile salt as is afforded us by wood, when that is first by an open fire divided into ashes and soot, and that soot is afterwards placed in a strong retort, and compelled by an urgent fire to part with its spirit, oil, and salt; for though I dare not peremptorily deny that in the liquors of guaiacum and other woods distilled in retorts after the common manner, there may be saline parts, which by reason of the analogy may pretend to the name of some kinde of volatile salts, yet questionless there is a great disparity betwixt such salts and that which we have sometimes obtained upon the first distillation of soot (though for the most part it has not been separated from the first or second rectification, and sometimes not till the third).   For we could never yet see separated from woods analysed only the vulgar way in close vessels any volatile salt in a dry and saline form, as that of soot, which we have often had very crystalline and geometrically figured.   And then, whereas the saline parts of the spirits of guaiacum, etc., appear upon distillation sluggish enough, the salt of soot seems to be one of the most volatile bodies in all nature; and if it be well made will readily ascend with the milde heat of a furnace, warmed only by the single wick of a lamp, to the top of the highest glass vessels that are commonly made use of for distillation: and besides all this, the taste and smell of the salt of soot are exceedingly differing from those of the spirits of guaiacum, etc., and the former not only smells and tastes much less like a vegetable salt, than like that of harts-horn, and other animal concretes, but in divers other properties seems more of kin to the family of animals than to that of vegetable salts, as I may elsewhere (God permitting) have an occasion more particularly to declare.   I might likewise by some other examples manifest that the chymists, to have dealt clearly, ought to have more explicitly and particularly declared by what degree of fire, and in what manner of application of it, they would have us judge a division made by the fire to be a true analysis into their principles, and the productions of it to deserve the name of elementary bodies.   But it is time that I proceed to mention the particular reasons that incline me to doubt whether the fire be the true and universal analyser of mixt bodies; of which reasons what has been already objected may pass for one.

In the next place I observe, that there are some mixt bodies from which it has not been yet made appear that any degree of fire can separate either salt or sulphur or mercury, much less all the three.   The most obvious instance of this truth is gold, which is a body so fixt, and wherein the elementary ingredients (if it have any) are so firmly united to each other, that we finde not in the operations wherein gold is exposed to the fire, how violent soever, that it does discernably so much as lose of its fixedness or weight, so far is it from being dissipated into those principles, whereof one at least is acknowledged to be fugitive enough; and so justly did the spagirical poet somewhere exclaim:

Cuncta adeo miris compagibus hærent.

And I must not omit on this occasion to mention to you, Eleutherius, the memorable experiment that I remember I met with in [1] Gasto Claveus, who, though a lawyer by

[1] Gasto Claveus *Apolog. Argur. and Chryfopera.*

profession, seems to have had no small curiosity and experience in chymical affairs: he relates then, that having put into one small earthen vessel an ounce of the most pure gold, and into another the like weight of pure silver, he placed them both in that part of a glass-house furnace wherein the workmen keep their metal (as our English artificers call their liquid glass) continually melted, and that having there kept both the gold and the silver in constant fusion for two months together, he afterwards took them out of the furnace and the vessels, and weighing both of them again, found that the silver had not lost above a twelfth part of its weight, but the gold had not of his lost anything at all. And though our author endeavours to give us of this a scholastic reason, which I suppose you would be as little satisfied with, as I was when I read it, yet for the matter of fact, which will serve our present turne, he assures us, that though it be strange, yet experience itself taught it him to be most true.

And though there be not perhaps any other body to be found so perfectly fixt as gold, yet there are divers others so fixt or composed, at least of so strictly united parts, that I have not yet observed the fire to separate from them any one of the chymist's principles. I need not tell you what complaints the more candid and judicious of the chymists themselves are wont to make of those boasters that confidently pretend, that they have extracted the salt or sulphur of quicksilver, when they have disguised it by additaments, wherewith it resembles the concretes whose names are given it; whereas by a skilful and rigid *examen*, it may be easily enough stript of its disguises, and made to appear again in the pristine form of running mercury. The pretended salts and sulphurs being so far from being elementary parts extracted out of the bodie of mercurie, that they are rather (to borrow a terme of the grammarians) de-compound bodies, made up of the whole metal and the menstruum, or other additaments imployed to disguise it. And as for silver, I never could see any degree of fire make it part with any of its three principles. And though the experiment lately mentioned from Claveus may beget a suspition that silver may be dissipated by fire, provided it be extreamly violent and very lasting, yet it will not necessarily follow, that because the fire was able at length to make the silver lose a little of its weight, it was therefore able to dissipate it into its principles. For first I might alledge that I have observed little grains of silver to lie hid in the small cavities (perhaps glassed over by a vitrifying heat) in crucibles, wherein silver has been long kept in fusion, whence some goldsmiths of my acquaintance make a benefit by grinding such crucibles to powder, to recover out of them the latent particles of silver. And hence I might argue, that perhaps Claveus was mistaken, and imagined that silver to have been driven away by the fire, that indeed lay in minute parts hid in his crucible, in whose pores so small a quantity as he misst of so ponderous a bodie might very well lie concealed.

But secondly, admitting that some parts of the silver were driven away by the violence of the fire, what proof is there that it was either the salt, the sulphur, or the mercury of the metal, and not rather a part of it homogeneous to what remained? For besides that the silver that was left seemed not sensibly altered, which probably would have appeared, had so much of any one of its principles been separated from it; we finde in other mineral bodies of a less permanent nature than silver, that the fire may divide them into such minute parts, as to be able to carry them away with itself, without at all destroying their nature. Thus we see that in the refining of silver, the lead that is mixt with it (to carry away the copper or other ignoble mineral that embases the silver) will, if it be let alone, in time evaporate away upon the test; but if (as is most usual amongst those that refine great quantities of metals together) the lead be blown off from the silver by bellowes, that which would else have gone away in the form of unheeded steams will in great part be collected not far from the silver, in the form of a darkish powder or calx; which, because it is blown off from silver, they call litharge of silver. And thus Agricola in divers places informs us, when copper, or the ore of it, is colliquated by the violence of the fire with cadmia, the sparks, that in great multitudes do fly upwards, do some of them stick to the vaulted roofs of the furnaces, in the form of little and (for the most part) white bubbles, which there- fore the Greeks, and, in imitation of them, our drugsters call *pompholyx :* and others more heavy partly adhere to the sides of the furnace, and partly (especially if the covers be not kept upon the pots) fall to the ground, and by reason of their ashy colour as well as weight were called by the same Greeks σποδòs, which, I need not tell you, in their language signifies ashes. I might add, that I have not found that from Venetian talc (I say Venetian because I have found other kinds of that mineral more open), from the *lapis ossifragus* (which the shops call *ostiocolla*), from Muscovia glass, from pure and fusible sand (to mention now no other concretes), those of my acquaintance that have tried, have been able by the fire to separate any one of the hypostatical principles; which you will the less scruple to believe, if you consider that glass may be made by the bare colliquation of the salt and earth remaining in the ashes of a burnt plant, and that yet common glass, once made, does so far resist the violence of the fire, that most chymists think it a body more undestroyable than gold itself. For if the artificer can so firmly unite such comparative gross particles as those of earth and salt that make up common ashes, into a body indissoluble by fire, why may not nature associate in divers bodies the more minute elementary corpuscles she has at hand too firmly to let them be separable by the fire? And on this occasion, Eleutherius, give me leave to mention to you two or three slight experiments, which will, I hope, be found more pertinent to our present discourse, than at first perhaps they will appear. The first is, that, having (for trial's sake) put a quantity of that fugitive concrete, camphire, into a glass vessel, and placed it in a gentle heat, I found it (not leaving behinde, according to my estimate, not so much as one grain) to sublime to the top of the vessel into flowers; which is whiteness, smell, etc., seemed not to differ from the cam- phire itself. Another experiment is that of Helmont, who in several places affirms, that a coal kept in a glass exactly closed will never be calcined to ashes, though kept never so long in a strong fire: to countenance which I shall tell you this trial of my own, that having sometimes distilled some woods, as particularly box, whilst our *caput mortuum* remained in the retort, it continued black like charcoal, though the retort were earthen, and kept red-hot in a vehement fire; but as soon as ever it was brought out of the candent vessel into the open air, the burning coals did hastily degenerate or fall asunder, without the assistance of any new calcination, into pure white ashes. And to these two I shall add but this obvious and known observation, that common sulphur (if it be pure and freed from its vinegar) being leasurely sublimed in close vessels, rises into dry flowers, which may be presently melted into a bodie of the same nature with that which afforded them. Though, if brimstone be burnt in the open air, it gives, you know, a penetrating fume, which being caught in a glass bell condenses into that acid liquor called oil of sulphur *per campanam.* The use I would make of these experiments collated with what I lately told you out of Agricola is this, that even among the bodies that are not fixt, there are divers of such a texture, that it will be hard to make it appear how the fire, as chymists are wont to imploy it, can resolve them into elementary substances. For some bodies being of such a texture that the fire can drive them into the cooler and less hot part of the vessels wherein they are included, and if need be, remove them from place to place to fly the greatest heat, more easily than it can divorce their elements (especially without the assistance of the air), we see that our chymists cannot analyse them in close vessels, and of other compound bodies the open fire can as little separate the elements. For what can a naked fire do to analyse a mixt bodie, if its component principles be so minute, and so strictly united, that the corpuscles of it need less heat to carry them up than is requisite to divide them into their principles? So that of some bodies the fire cannot in close vessels make any analysis at all; and others will in the open air fly away in the forms of flowers or liquors, before the heat can prove able to divide them into their principles. And this may hold, whether the various similar parts of a concrete be combined by nature or by art; for in factitious sal ammoniac we finde the common and the urinous salts so well mingled, that both in the open fire, and in subliming vessels they rise together as one salt, which seems in such vessels irresoluble by fire alone. For I can shew you sal ammoniac which after the ninth sublimation does still retain its compounded nature. And indeed I scarce know any one mineral, from which by fire alone chymists are wont to sever any substance simple enough to deserve the name of an element or principle. For though out of native cinnaber they distil quicksilver, and though from many of those stones that the ancients called pyrites they sublime brimstone, yet both that quicksilver and this sulphur being very often the same with the common minerals that are sold in the shops under those names, are themselves too much compounded bodies to pass for the elements of such. And thus much, Eleutherius, for the second argument that belongs to my first consideration; the others I shall the lesse insist on, because I have dwelt so long upon this.

Proceed we then in the next place to consider, that there are divers separations to be made by other means, which either cannot at all, or else cannot so well be made by the fire alone. When gold and silver are melted into one mass, it would lay a great obligation upon refiners and goldsmiths to teach them the art of separating them by the fire, without the trouble and charge they are fain to be at to sever them. Whereas they may be very easily parted by the affusion of spirit of nitre or *aqua fortis ;* which the French therefore call *eau de depart :* so likewise the metalline part of vitriol will not be so easily and conveniently separated from the saline part even by a violent fire, as by the affusion of certain alkalisate salts in a liquid form upon the solution of vitriol made in common water. For thereby the acid salt of the vitriol leaving the copper it had corroded to join with the added salts, the metalline part will be precipitated to the bottom almost like mud. And that I may not give instances only in de-compound bodies, I will add a not useless one of another kinde. Not only chymists have not been able (for ought is vulgarly known) by fire alone to separate true sulphur from antimony, but though you may finde in their books many plausible processes of extracting it, yet he that shall make as many fruitless trials as I have done to obtain it by, most of them will, I suppose, be easily persuaded, that the productions of such processes are antimonial sulphurs rather in name than nature. But though antimony sublimed by itself is reduced but to a volatile powder, or antimonial flowers, of a compounded nature like the mineral that affords them: yet I remember that some years ago I sublimed out of antimony a sulphur, and that in greater plenty than ever I saw obtained from that mineral, by a method which I shall therefore acquaint you with, because chymists seem not to have taken notice of what importance such experiments may be in the indagation of the nature, and especially of the number of the elements. Having then purposely for trial's sake digested eight ounces of good and well powdered antimony with twelve ounces of oil of vitriol in a well stopt glass vessel for about six or seven weeks; and having caused the mass (grown hard and brittle) to be distilled in a retort placed in sand, with a strong fire; we found the antimony to be so opened, or altered by the *menstruum* wherewith it had been digested, that whereas crude antimony, forced up by the fire, arises only in flowers, our antimony thus handled afforded us partly in the receiver, and partly in the neck and at the top of the retort, about an ounce of sulphur, yellow and brittle like common brimstone, and of so sulphureous a smell, that upon the unluting the vessels it infected the room with a scarce supportable stink. And this sulphur, besides the colour and smell, had the perfect inflammability of common brimstone, and would immediately kindle (at the flame of a candle) and burn blue like it. And though it seemed that the long digestion wherein our antimony and *menstruum* were detained, did conduce to the better unlocking of the mineral, yet if you have not the leasure to make so long a digestion you may by incorporating with powdered antimony a convenient quantity of oil of vitriol, and committing them immediately to distillation, obtain a little sulphur like unto the common one, and more combustible than perhaps you will at first take notice of. For I have observed, that though (after its being first kindled) the flame would sometimes go out too soon of itself, if the same lump of sulphur were held again to the flame of a candle, it would be rekindled and burn a pretty while, not only after the second, but after the third or fourth accension. You, to whom I think I shewed my way of discovering something of sulphureous in oil of vitriol, may perchance suspect, Eleutherius, either that this substance was some venereal sulphur that lay hid in that liquor, and was by this operation only reduced into a manifest body; or else that it was a compound of the unctuous parts of the antimony, and the saline ones of the vitriol, in regard that (as Gunther informs us) divers learned men would have sulphur to be nothing but a mixture made in the bowels of the earth of vitriolate spirits and a certain combustible substance. But the quantity of sulphur we obtained by digestion was much too great to have been latent in the oil of vitriol. And that vitriolate spirits are not necessary to the construction of such a sulphur as ours, I could easily manifest, if I would acquaint you with the several wayes by which I have obtained, though not in such plenty, a sulphur of antimony, coloured and combustible like common brimstone. And though I am not now minded to discover them, yet I shall tell you, that to satisfie some ingenious men, that distilled vitriolate spirits are not necessary to the obtaining of such a sulphur as we have been considering, I did by the bare distillation of only spirit of nitre, from its weight of crude antimony separate, in a short time, a yellow and very inflammable sulphur, which, for ought I know, deserves as much the name of an element as anything that chymists are wont to separate from any mineral by the fire. I could perhaps tell you of other operations upon antimony, whereby that may be extracted from it, which cannot be forced out of it by the fire; but I shall reserve them for a fitter opportunity, and only annex at present this slight, but not impertinent experiment. That whereas I lately observed to you, that the urinous and common salts whereof sal ammoniac consists, remained unsevered by the fire in many successive sublimations, they may be easily separated, and partly without any fire at all, by pouring upon the concrete finely powdered, a solution of salt of tartar, or of the salt of wood-ashes; for upon your diligently mixing of these you will finde your nose invaded with a very strong smell of urine, and perhaps too your eyes forced to water, by the same subtle and piercing body that produces the stink; both these effects proceeding from hence, that by the alkalisate salt, the sea salt that entered the composition of the sal ammoniac is mortified and made more fixt, and thereby a divorce is made between it and the volatile urinous salt, which being at once set at liberty, and put into motion, begins presently to fly away, and to offend the nostrils and eyes it meets with by the way. And if the operation of these salts be in convenient glasses promoted by warmth, though but by that of a bath, the ascending steames may easily be caught and reduced into a penetrant spirit, abounding with a salt, which I have sometimes found to be separable in a crystalline form. I might add to these instances, that where as sublimate, consisting, as you know, of salts and quicksilver combined and carried up together by heat, may be sublimed, I know not how often, by a like degree of fire, without suffering any divorce of the component bodies, the mercury may be easily severed from the adhering salts, if the sublimate be distilled from salt of tartar, quicklime, or such alkalisate bodies. But I will rather observe to you, Eleutherius, what divers ingenious men have thought somewhat strange, that by such an additament that seems but only to promote the separation, there may be easily obtained from a concrete, that by the fire alone is easily divisible into all the elements that vegetables are supposed to consist of, such a similar substance as differs in many respects from them all, and consequently has by many of the most intelligent chymists been denied to be contained in the mixt body. For I know a way, and have practised it, whereby common tartar, without the addition of anything that is not perfectly a mineral, except saltpetre, may by one distillation in an earthen retort be made to afford good store of real salt, readily dissoluble in water, which I found to be neither acid, nor of the smell of tartar, and to be almost as volatile as spirit of wine itself, and to be indeed of so differing a nature from all that is wont to be separated by fire from tartar, and divers learned men, with whom I discoursed of it, could hardly be brought to believe, that so fugitive a salt could be afforded by tartar, till I assured it them upon my own knowledge. And if I did not think you apt to suspect me to be rather too backward than too forward to credit or affirm unlikely things, I could convince you by what I have yet lying by me of that anomalous salt.

The fourth thing that I shall alledge to countenance my first consideration is, that the fire even when it divides a body into substances of divers consistences, does not most commonly analyse it into hypostatical principles, but only disposes its parts into new textures, and thereby produces concretes of a new indeed, but yet of a compound nature. This argument it will be requisite for me to prosecute so fully hereafter, that I hope you will then confess that 'tis not for want of good proofs that I desire leave to suspend my proofs till the series of my discourse shall make it more proper and seasonable to propose them.

It may be further alledged on the behalf of my first consideration, that some such distinct substances may be obtained from some concretes without fire, as deserve no less the name of elementary than many that chymists extort by the violence of the fire.

We see that the inflammable spirit, or as the chymists esteem it, the sulphur of wine, may not only be separated from it by the gentle heat of a bath, but may be distilled either by the help of the sunbeams, or even of a dunghill, being indeed of so fugitive a nature, that it is not easy to keep it from flying away, even without the application of external heat. I have likewise observed that a vessel full of urine being placed in a dunghill, the putrefaction is wont after some weeks so to open the body, that the parts disbanding the saline spirit, will within no very long time, if the vessel be not stoppt, fly away of itself; insomuch that from such urine I have been able to distil little or nothing else than a nauseous phlegme, instead of the active and piercing salt and spirit that it would have afforded, when first exposed to the fire, if the vessel had been carefully stoppt.

And this leads me to consider, in the fifth place, that it will be very hard to prove, that there can no other body or way be given which will as well as the fire divide concretes into several homogeneous substances, which may consequently be called their elements or principles, as well as those separated or produced by the fire. For since we have lately seen, that nature can successfully employ other instruments than the fire to separate distinct substances from mixt bodies, how know we, but that nature has made, or art may make, some such substance as may be a fit instrument to analyse mixt bodies, or that some such method may be found by human industry or luck, by whose means compound bodies may be resolved into other substances than such as they are wont to be divided into by the fire. And why the products of such an analysis may not as justly be called the component principles of the bodies that afford them, it will not be easy to shew, especially since I shall hereafter make it evident, that the substances which chymists are wont to call the salts, and sulphurs, and mercuries of bodies, are not so pure and elementary as they presume, and as their hypothesis requires. And this may therefore be the more freely pressed upon the chymists, because neither the Paracelsansi, nor the Helmontians can reject it without apparent injury to their respective masters. For Helmont does more than once inform his readers, that both Paracelsus and himself were possessors of the famous liquor, alkahest, which for its great power in resolving bodies irresoluble by vulgar fires, he somewhere seems to call *ignis Gehennæ*. To this liquor he ascribes (and that in great part upon his own experience) such wonders, that if we suppose them all true, I am so much the more a friend to knowledge than to wealth, that I should think the alkahest a nobler and more desirable secret than the philosopher's stone itself. Of this universal dissolvent he relates, that having digested with it for a compentet time a piece of oaken charcoal, it was thereby reduced into a couple of new and distinct liquors, discriminated from each other by their colour and situation, and that the whole body of the coal was reduced into those liquors, both of them separable from his immortal menstruum, which remained as fit for such operations as before. And he moreover tells us in divers places of his writings, that by his powerful, and unwearied agent, he could dissolve metals, marchasites, stones, vegetable and animal bodies of what kinde soever, and even glass itself (first reduced to powder), and in a word, all kind of mixt bodies in the world, into their several similar substances, without any residence or *caput mortuum*. And lastly, we may gather this further from his informations, that the homogeneous substances obtainable from compound bodies by his piercing liquor, were oftentimes different enough, both as to number and as to nature, from those into which the same bodies are wont to be divided by common fire. Of which I shall need in this place to mention no other proof, than what whereas we know that in our common analysis of a mixt body there remains a terrestrial and very fixt substance, oftentimes associated with a salt as fixt; our author tells us, that by his way he could distil over all concretes without any *caput mortuum*, and consequently could make those parts of the concrete volatile, which in the vulgar analysis would have been fixt. So that if our chymists will not reject the solemn and repeated testimony of a person, who cannot but be acknowledged for one of the greatest spagyrists that they can boast of, they must not deny that there is to be found in nature another agent able to analyse compound bodies less violently, and both more genuinely and more universally than the fire. And for my own part, though I cannot but say on this occasion what (you know) our friend Mr. Boyle is wont to say, when he is askt his opinion of any strange experiment; *That he that hath seen it hath more reason to believe it, than he that hath not*, yet I have found Helmont so faithful a writer, even in divers of his improbable experiments (I alwaies except that extravagant treatise *De Magnetica Vulnerum Curatione*, which some of his friends affirm to have been first published by his enemies) that I think it somewhat harsh to give him the lye, especially to what he delivers upon his own proper tryal. And I have heard from very credible eye-witnesses some things, and seen some others myself, which argue so strongly, that a circulated salt, or a menstruum (such as it may be) may by being abstracted from compound bodies, whether mineral, animal, or vegetable, leave them more unlockt than a wary naturalist would easily believe, that I dare not confidently measure the power of nature and art by that of tne menstruums, and other instruments that eminent chymists themselves are as yet wont to employ about the analysing of bodies; nor deny that a menstruum may at least from this or that particular concrete obtain some apparently similar substance, differing from any obtainable from the same body by any degree or manner of application of the fire. And I am the more backward to deny peremptorily, that there may be such openers of compound bodies, because among the experiments that make me speak thus warily, there wanted not some in which it appeared not, that one of the substances, not separable by common fires and menstruums, could retain anything of the salt by which the separation was made.

And here, Eleutherius (says Carneades) I should conclude as much of my discourse as belongs to the first consideration I proposed, but that I foresee, that what I have delivered will appear liable to two such specious objections, that I cannot safely proceed any further till I have examined them.

And first, one sort of opposers will be forward to tell me, that they do not pretend by fire alone to separate out of all compound bodies their *hypostatical* principles; it being sufficient that the fire divides them into such, though afterwards they employ other bodies to collect the similar parts of the compound; as 'tis known, that though they make use of water to collect the saline parts of ashes from the terrestrial wherewith they are blended, yet it is the fire only that incinerates bodies, and reduces the fixed part of them into the salt and earth, whereof ashes are made up. This objection is not, I confess, inconsiderable, and I might in great part allow of it, without granting it to make against me, if I would content myself to answer, that it is not against those that make it that I have been disputing, but against those vulgar chymists, who themselves believe, and would fain make others do so, that the fire is not only an universal, but an adequate and sufficient instrument to analyse mixt bodies with. For as to their practice of extracting the fixed salt out of ashes by the affusion of water, 'tis obvious to alledge, that the water does only assemble together the salt, the fire had before divided from the earth: as a sieve does not further break the corn, but only bring together into two distinct heaps the flower and the bran, whose corpuscles before lay promiscuously blended together in the meal. This I say I might alledge, and thereby exempt myself from the need of taking any farther notice of the proposed objection. But not to lose the rise it may afford me of illustrating the matter under consideration, I am content briefly to consider it, as far forth as my present disquisition may be concerned in it.

Not to repeat then what has been already answered, I say further, that though I am so civil an adversary, that I will allow the chymists, after the fire has done all its work, the use of fair water to make their extractions with, in such cases wherein the water does not co-operate with the fire to make the analysis; yet since I grant this but upon supposition that the water does only wash off the saline particles, which the fire alone has before extricated in the analysed body, it will not be reasonable, that this concession should extend to other liquors that may add to what they dissolve, nor so much as to other cases than those newly mentioned: which limitation I desire you would be pleased to bear in mind till I shall anon have occasion to make use of it. And this being thus premised, I shall proceed to observe,

First, that many of the instances I proposed in the

preceding discourse are such, that the objection we are considering will not at all reach them. For fire can no more with the assistance of water, than without it, separate any of the three principles, either from gold, silver, mercury, or some others of the concretes named above.

Hence we may inferre, that fire is not an universal analyser of all mixt bodies, since of metals and minerals, wherein chymists have most exercised themselves, there appear scarce any which they are able to analyse by fire, nay, from which they can unquestionably separate so much as any one of their hypostatical principles; which may well appear no small disparagement, as well to their hypothesis, as to their pretensions.

It will also remain true, notwithstanding the objection, that there may be other wayes, than the wonted analysis by fire, to separate from a compound body substances as homogeneous as those that chymists scruple not to reckon among their *tria prima* (as some of them, for brevity sake, call their three principles).

And it appears, that by convenient additaments such substances may be separated by the help of the fire, as could not be so by the fire alone. Witness the sulphur of antimony.

And lastly, I must represent, that since it appears too that the fire is but one of the instruments that must be employed in the resolution of bodies, we may reasonably challenge the liberty of doing two things. For whenever any menstruum or other additament is employed, together with the fire to obtain a sulphur or a salt from a body, we may well take the freedom to examine, whether or no that menstruum do barely help to separate the principle obtained by it, or whether there intervene not a coalition of the parts of the body wrought upon with those of the menstruum, whereby the produced concrete may be judged to result from the union of both. And it will be farther allowable for us to consider, how far any substance, separated by the help of such additaments, ought to pass for one of the *tria prima ;* since by one way of handling the same mixt body, it may, according to the nature of the additaments, and the method of working upon it, be made to afford differing substances from those obtainable from it by other additaments, and another method, nay and (as may appear by what I formerly told you about tartar) differing from any of the substances into which a concrete is divisible by the fire without additaments, though perhaps those additaments do not, as ingredients, enter the composition of the obtained body, but only diversify the operation of the fire upon the concrete; and though that concrete by the fire alone may be divided into a number of differing substances, as great as any of the chymists, that I have met with, teach us that of the elements to be. And having said thus much (saies Carneades) to the objection likely to be proposed by some chymists, I am now to examine that which I foresee will be confidently pressed by divers peripateticks, who, to prove fire to be the true analyser of bodies, will plead, that it is the very definition of heat given by Aristotle, and generally received, *congregare homogenea, et heterogenea segregare,* to assemble things of a resembling, and disjoyn those of a differing nature. To this I answer, that this effect is far from being so essential to heat, as 'tis generally imagined; for it rather seems, that the true and genuine property of heat is, to set a moving, and thereby to dissociate the parts of bodies, and subdivide them into minute particles, without regard to their being homogeneous or heterogeneous, as is apparent in the boyling of water, the distillation of quicksilver, or the exposing of bodies to the action of the fire, whose parts either are not (at least in that degree of heat appear not) dissimilar, where, all that the fire can do, is to divide the body into very minute parts which are of the same nature with one another, and with their *totum,* as their reduction by condensation evinces. And even when the fire seems most so *congregare homogenea, et segregare heterogenea,* it produces that effect but by accident; for the fire does but dissolve the cement, or rather shatter the frame, or structure that kept the heterogeneous parts of bodies together, under one common form; upon which dissolution the component particles of the mixt, being freed and set at liberty, do naturally, and oftentimes without any operation of the fire, associate themselves each with its like, or rather do take those places which their several degrees of gravity and levity, fixedness or volatility (either natural, or adventitious from the impression of the fire) assigne them. Thus in the distillation (for instance) of man's blood, the fire does first begin to dissolve the *nexus* or cement of the body; and then the water, being the most volatile, and easy to be extracted, is either by the igneous atomes, or the agitation they are put into by the fire, first carried up, till forsaken by what carried it up, its weight sinks it down, into the receiver: but all this while the other principles of the concrete remain unsevered, and require a stronger degree of heat to make a separation of its more fixt elements; and therefore the fire must be increased which carries over the volatile salt and the spirit, they being, though believed to be differing principles, and though really of different consistency, yet of an almost equal volatility. After them, as less fugitive, comes over the oyl, and leaves behinde the earth and the alcali, which being of an equal fixednesse, the fire severs them not, for all the definition of the schools. And if into a red-hot earthen or iron retort you cast the matter to be distilled, you may observe, as I have often done, that the predominant fire will carry up all the volatile elements confusedly in one fume, which will afterwards take their places in the receiver, either according to the degree of their gravity, or according to the exigency of their respective textures; the salt adhering, for the most part, to the sides and top, and the phlegme fastening itself there too in great drops, the oyle and spirit placing themselves under, or above one another, according as their ponderousness makes them swim or sink. For 'tis observable, that though oyl or liquid sulphur be one of the elements separated by this fiery analysis, yet the heat which accidentally unites the particles of the other volatile principles, has not always the same operation on this, there being divers bodies which yield two oyls, whereof the one sinks to the bottom of that spirit on which the other swims; as I can shew you in some oyls of the same deers blood, which are yet by me; nay I can shew you two oyls carefully made of the same parcel of humane blood, which not only differ extreamly in colour, but swim upon one another without mixture, and if by agitation confounded will of themselves divorce again.

And that the fire doth oftentimes divide bodies, upon the account that some of their parts are more fixt, and some more volatile, how far soever either of these two may be from a pure elementary nature is obvious enough, if men would but heed it in the burning of wood, which the fire dissipates into smoake and ashes: for not only the latter of these is confessedly made up of two such differing bodies as earth and salt; but the former being condensed into that soot which adheres to our chimneys, discovers itself to contain both salt and oyl, and spirit and earth, (and some portion of phlegme too) which being, all almost, equally volatile to that degree of fire which forces them up, (the more volatile parts helping perhaps, as well as the urgency of the fire, to carry up the more fixt ones, as I have often tried in dulcified colcothar, sublimed by sal amoniack blended with it) are carried up together, but may afterwards be separated by other degrees of fire, whose orderly gradation allowes the disparity of their volatileness to discover itself. Besides, if differing bodies united into one mass be both sufficiently fixt, the fire finding no parts volatile enough to be expelled or carried up, makes no separation at all; as may appear by a mixture of colliquated silver and gold, whose component metals may be easily severed by *aqua fortis*, or *aqua regis* (according to the predominancy of the silver or the gold) but in the fire alone, though vehement, the metals remain unsevered, the fire only dividing the body into smaller particles (whose littleness may be argued from their fluidity) in which either the little nimble atoms of fire, or its brisk and numberless strokes upon the vessels, hinder rest and continuity, without any sequestration of elementary principles. Moreover, the fire sometimes does not separate, so much as unite, bodies of a differing nature; provided they be of an almost resembling fixedness, and have in the figure of their parts an aptness to coalition, as we see in the making of many plaisters, oyntments, etc. And in such metalline mixtures as that made by melting together two parts of clean brass with one of pure copper, of which some ingenious tradesmen cast such curious patterns (for gold and silver works) as I have sometimes taken great pleasure to look upon. Sometimes the bodies mingeld by the fire are differing enough as to fixidity and volatility, and yet are so combined by the first operation of the fire, that itself does scarce afterwards separate them, but only pulverise them; whereof an instance is afforded us by the common preparation of *mercurius dulcis*, where the saline particles of the vitriol, sea salt, and sometimes nitre, employed to make the sublimate, do so unite themselves with the mercurial particles made use of, first to make sublimate, and then to dulcifie it, that the saline and metalline parts arise together in many successive sublimations, as if they all made but one body. And sometimes too the fire does not only not sever the differing elements of a body, but combine them so firmly, that nature herself does very seldom, if ever, make unions less dissoluble. For the fire meeting with some bodies exceedingly and almost equally fixt, instead of making a separation, makes an union so strict, that itself, alone, is unable to dissolve it; as we see, when an alcalisate salt and the terrestrial residue of the ashes are incorporated with pure sand, and by vitrification made one permanent body (I mean the course or greenish sort of glass) that mocks the greatest violence of the fire, which though able to marry the ingredients of it, yet is not able to divorce them. I can shew you some pieces of glass which I saw flow down from an earthen crucible purposely exposed for a good while, with silver in it, to a very vehement fire. And some that deal much in the fusion of metals informe me, that the melting of a great part of a crucible into glass is no great wonder in their furnaces. I remember I have observed too in the melting of great quantities of iron out of the oar, by the help of store of charcoal (for they affirm that sea-coal will not yield a flame strong enough) that by the prodigious vehemence of the fire, excited by vast bellows (made to play by great wheels turned about by water) part of the materials exposed to it was, instead of being analysed, colliquated, and turned into a dark, solid and very ponderous glass, and that in such quantity, that in some places I have seen the very highwayes, neer such iron-works, mended with heaps of such lumps of glasse, instead of stones and gravel. And I have also observed, that some kind of fire-stone itself, having been employed in furnaces wherein it was exposed to very strong and lasting fires, has had all its fixt parts so wrought on by the fire, as to be perfectly vitrified, which I have tried by forcing from it pretty large pieces of perfect and transparent glass. And lest you might think, Eleutherius, that the questioned definition of heat may be demonstrated, by the definition which is wont to be given and acquiesced in, of its contrary quality, cold, whose property is taught to be *tam honogenea, quam heterogenea congregare*, give me leave to represent to you, that neither is this definition unquestionable; for not to mention the exceptions, which a logician, as such, may take at it, I consider that the union of heterogeneous bodies which is supposed to be the genuine production of cold, is not performed by every degree of cold. For we see for instance that in the urine of healthy men, when the liquor has been suffered a while to stand, the cold makes a separation of the thinner part from the grosser, which subsides to the bottom, and growes opacous there; whereas if the urinal be warme, these parts readily mingle again, and the whole liquor becomes transparent as before. And when, by glaciation, wood, straw, dust, water, etc. are supposed to be united into one lump of ice, the cold does not cause any real union or adunation (if I may so speak) of these bodies, but only hardening the aqueous parts of the liquor into ice, the other bodies being accidentally present in that liquor are frozen up in it, but not really united. And accordingly if we expose a heap of mony consisting of gold, silver and copper coynes, or any other bodies of differing natures, which are destitute of aqueous moisture, capable of congelation, to never so intense a cold, we find not that these differing bodies are at all thereby so much as compacted, much less united together; and even in liquors themselves we find phænomena which induce us to question the definition which we are examining. If Paracelsus his authority were to be looked upon as a sufficient proof in matters of this nature, I might here insist on that process of his, whereby he teaches that the essence of wine may be severed from the phlegme and ignoble part by the assistance of congelation: and because much weight has been laid upon this process, not only by Paracelsians, but other writers, some of whom seem not to have perused it themselves, I shall give you the entire passage in the author's own words, as I lately found them in the sixth book of his *Archidoxis*, an extract whereof I have yet about me; and it sounds thus. " De vino sciendum est, fæcem phlegmaque ejus esse mineram, et vini substantiam esse corpus in quo conservatur essentia, prout auri in auro latet essentia. Juxta quod practicam nobis ad memoriam ponimus, ut non obliviscamur, ad hunc modum: recipe vinum vetustissimum et optimum quod hahere poteris, calore saporeque ad placitum, hoc in vas vitreum infundas ut tertiam ejus partem impleat, et sigillo hermetis occlusum in equino ventre mensibus quatuor, et in continuato calore teneatur qui non deficiat. Quo peracto, hyeme cum frigus et gelu maxime sæviunt, his per mensem exponatur ut congeletur. Ad hunc modum frigus vini spiritum una cum ejus substantia protrudit in vini centrum, ac separat a phlegmate: conge-latum abjice, quod vero congelatum non est, id spiritum cum substantia esse judicato. Hunc in pelicanum positum in arenæ digestione non adeo calida per aliquod tempus manere sinito; postmodum eximito vini magis-terium, de quo locuti sumus."

But I dare not Eleu. lay much weight upon this process, because I have found that if it were true, it would be but seldom practicable in this countrey upon the best wine: for though this present winter hath been extra-ordinary cold, yet in very keen frosts accompanied with lasting snowes, I have not been able in any measure to freez a thin vial full of sack; and even with snow and salt I could freeze little more than the surface of it; and I suppose Eleu. that 'tis not every degree of cold that is capable of congealing liquors, which is able to make such an analysis (if I may so call it) of them by separating their aqueous and spirituous parts; for I have sometimes, though not often, frozen severally, red-wine, urine and milk, but could not observe the expected separation. And the Dutchmen that were forced to winter in that icie region neer the artick circle, called Nova Zembla, although they relate, as we shall see below, that there was a separation of parts made in their frozen beer about the middle of November, yet of the freezing of their sack in December following they give but this account: " Yea and our sack, which is so hot, was frozen very hard, so that when we were every man to have his part, we were forced to melt it in the fire; which we shared every second day, about half a pinte for a man, wherewith we were forced to sustain ourselves." In which words they imply not, that their sack was divided by the frost into differing substances, after such manner as their beer had been. All which notwithstanding, Eleu. suppose that it may be made to appear, that even cold sometimes may *congregare homogenea, et heteroghnea segregare:* and to manifest this I may tell you, that I did once, purposely, cause to be decocted in fair water a plant abounding with sulphureous and spirituous parts, and having exposed the decoction to a keen north-wind in a very frosty night, I observed, that the more aqueous parts of it were turned by the next morning into ice, towards the innermost part of which, the more agile and spirituous parts, as I then conjectured, having retreated, to shun as much as might be their environing enemy, they had there preserved themselves unfrozen in the form of a high coloured liquor; the aqueous and spirituous parts having been so slightly (blended rather than) united in the decoction, that they were easily separable by such a degree of cold, as would not have been able to have divorced the parts of urine or wine, which by fermentation or digestion are wont, as tryal has informed me, to be more intimately associated each with other. But I have already intimated, Eleutherius, that I shall not insist on this experiment, not only because, having made it but once I may possibly have been mistaken in it; but also (and that principally) because of that much more full and eminent experiment of the separative vertue of extream cold, that was made, against their wills, by the forementioned Dutchmen that wintered in Nova Zembla; the relation of whose voyage being a very scarce book, it will not be amiss to give you that memorable part of it which concerns our present theme, as I caused the passage to be extracted out of the Englished voyage itself.

" Gerard de Veer, John Cornelyson and others, sent out of Amsterdam, anno dom. 1596, being forced by unseasonable weather to winter in Nova Zembla, near Ice-Haven; on the thirteenth of October, three of us (saies the relation) went aboard the ship, and laded a sled with beer; but when we had laden it, thinking to go to our house with it, suddenly there arose such a winde, and so great a storm and cold, that we were forced to go into the ship again, because we were not able to stay without; and we could not get the beer into the ship again, but were forced to let it stand without upon the sled: the fourteenth, as we came out of the ship, we found the barrel of beer standing upon the sled, but it was fast frozen at the heads; yet by reason of the great cold, the beer that purged out, froze as hard upon the side of the barrel, as if it had been glued thereon: and in that sort we drew it to our house, and set the barrel on end, and drank it up; but first we were forced to melt the beer, for there was scarce any unfrozen beer in the barrel; but in that thick yeast that was unfrozen, lay the strength of the beer, so that it was too strong to drink alone, and that which was frozen tasted like water; and being melted we mixed one with the other, and so drank it; but it had neither strength not taste."

And on this occasion I remember, that having the last very sharp winter purposely tried to freeze, among other liquors, some beer moderately strong, in glass vessels, with snow and salt, I observed, that there came out of the neck a certain thick substance, which, it seems, was much better able than the rest of the liquor (that I found turned into ice) to resist a frost; and which, by its colour and consistence seemed manifestly enough to be yeast, whereat, I confess, I somewhat marvelled, because I did not either discerne by the taste, or find by enquiry, that the beer was at all too new to be very fit to be drank. I might confirm the Dutchmen's relation, by what happened a while since to a neere friend of mine, who complained to me, that having brewed some beer or ale for his own drinking in Holland (where he then dwelt) the keenness of the late bitter winter froze the drink so as to reduce it into ice, and a small proportion of a very strong and spirituous liquor. But I must not entertaine you any longer concerning cold, not onely because you may think I have but lost my way into a theme which does not directly belong to my present undertaking; but because I have already enlarged myself too much upon the first consideration I proposed, though it appears so much a paradox, that it seemed to require that I should say much to keep it from being thought a meer extravagance; yet since I undertook but to make the common assumption of our chymists and Aristotelians appear questionable, I hope I have so performed that task, that I may now proceed to my following considerations, and insist less on them than I have done on the first.

THE SECOND PART

THE second consideration I desire to have notice taken of, is this; That it is not so sure, as both chymists and Aristotelians are wont to think it, that every seemingly similar or distinct substance that is separated from a body by the help of the fire, was pre-existent in it as a principle or element of it.

That I may not make this paradox a greater than I needs must, I will first briefly explain what the proposition means, before I proceed to argue for it.

And I suppose you will easily believe that I do not mean that anything is separable from a body by fire, that was not materially pre-existent in it; for it far exceeds the power of meerly naturall agents, and consequently of the fire, to produce anew, so much as one atome of matter, which they can but modifie and alter, not create; which is so obvious a truth, that almost all sects of philosophers have denied the power of producing matter to second causes; and the Epicureans and some others have done the like, in reference to their gods themselves.

Nor does the proposition peremptorily deny, but that some things obtained by the fire from a mixt body, may have been more than barely materially pre-existent in it, since there are concretes, which before they be exposed to the fire afford us several documents of their abounding, some with salt, and others with sulphur. For it will serve the present turn, if it appear that diverse things obtained from a mixt body exposed to the fire, were not its ingredients before: for if this be made to appear, it will be rationall enough to suspect that chymists may deceive themselves, and others, in concluding resolutely and universally, those substances to be the elementary ingredients of bodies barely separated by the fire, of which it yet may be doubted, whether there be such or no; at

63

least till some other argument, than that drawn from the analysis, be brought to resolve the doubt.

That then which I mean by the proposition I am explaining, is, that it may without absurdity be doubted whether or no the differing substances obtainable from a concrete dissipated by the fire were so existent in it in that forme (at least as to their minute parts) wherein we find them when the analysis is over, that the fire did only disjoyne and extricate the corpuscles of one principle from those of the other wherewith before they were blended.

Having thus explained my proposition, I shall endeavour to do two things, to prove it; the first of which is to shew that such substances as chymists call principles may be produced *de novo* (as they speak). And the other is to make it probable, that by the fire we may actually obtain from some mixt bodies such substances, as were not in the newly expounded sence, pre-existent in them.

To begin then with the first of these, I consider that if it be as true, as 'tis probable, that compounded bodies differ from one another but in the various textures resulting from the bigness, shape, motion, and contrivance of their small parts, it will not be irrational to conceive that one and the same parcel of the universall matter may by various alterations and contextures be brought to deserve the name, sometimes of a sulphureous, and sometimes of a terrene, or aqueous body. And this I could more largely explicate, but that our friend Mr. Boyle has promised us something about qualities, wherein the theme I now willingly resign him, will I question not be studiously enquired into. Wherefore what I shall now advance in favour of what I have lately delivered shall be deduced from experiments made divers years since. The first of which would have been much more considerable, but that by some intervening accidents I was necessitated to lose the best time of the year, for a trial of the nature of that I designed; it being about the middle of May before I was able to begin an experiment which should have then been two moneths old; but such as it was, it will not perhaps be impertinent to give you this narrative of it. At the time newly mentioned, I caused my gardiner (being by urgent occasions hindered from being present myself) to dig out a convenient quantity of good earth, and dry it well in an oven, to weigh it, to put it in an earthen pot almost level with the surface of the ground, and to set in it a selected seed he had before received from me, for that purpose, of squash, which is an Indian kind of pompion, that growes apace; this seed I ordered him to water only with rain or spring water. I did not (when my occasions permitted me to visit it) without delight behold how fast it grew, though unseasonably sown; but the hastning winter hindered it from attaining anything neer its due and wonted magnitude; (for I found the same autumn, in my garden, some of those plants, by measure, as big about as my middle) and made me order the having it taken up; which about the middle of October was carefully done by the same gardiner, who a while after sent me this account of it: " I have weighed the pompion with the stalk and leaves, all which weighed three pound wanting a quarter; then I took the earth, baked it as formerly, and found it just as much as I did at first, which made me think I had not dried it sufficiently: then I put it into the oven twice more, after the bread was drawn, and weighed it the second time, but found it shrink little or nothing."

But to deal candidly with you, Eleutherius, I must not conceal from you the event of another experiment of this kind made this present summer, wherein the earth seems to have been much more wasted; as may appear by the following account, lately sent me by the same gardiner, in these words. " To give you an account of your cucumbers, I have gained two indifferent fair ones, the weight of them is ten pound and a halfe, the branches with the roots weighed four pounds wanting two ounces; and when I had weighed them I took the earth, and baked it in several small earthen dishes in an oven; and when I had so done, I found the earth wanted a pound and a halfe of what it was formerly; yet I was not satisfied, doubting the earth was not dry: I put it into an oven the second time, (after the bread was drawn) and after I had taken it out and weighed it, I found it to be the same weight. So I suppose there was no moisture left in the earth. Neither do I think that the pound and half that was wanting was drawn away by the cucumber but a great part of it in the ordering was in dust (and the like) wasted: (the cucumbers are kept by themselves, lest you should send for them "). But yet in this tryal, Eleutherius, it appears that though some of the earth, or rather the dissoluble salt harboured in it, were wasted, the main body of the plant consisted of transmuted water. And I might add, that a year after I caused the formerly mentioned experiment, touching large pompions, to be reiterated, with so good success, that if my memory does not much misinform me, it did not only much surpass many that I made before, but seemed strangely to conclude what I am pleading for; though (by reason I have unhappily lost the particular account my gardiner writ me up of the circumstances) I dare not insist upon them. The like experiment may be as conveniently tried with the seeds of any plant, whose growth is hasty, and its size bulky. If tobacco will in these cold climates grow well in earth undunged, it would not be amiss to make a tryal with it; for 'tis an annual plant, that arises where it prospers, sometimes as high as a tall man, and I have had leaves of it in my garden neer a foot and a halfe broad. But the next time I try this experiment, it shall be with several seeds of the same sort, in the same pot of earth, that so the event may be the more conspicuous. But because everybody has not conveniency of time and place for this experiment neither, I made in my chamber, some shorter and more expeditious tryals. I took a top of spearmint, about an inch long, and put it into a good vial full of spring water, so as the upper part of the mint was above the neck of the glass, and the lower part immersed in the water; within a few dayes this mint began to shoot forth roots into the water, and to display its leaves, and aspire upwards; and in a short time it had numerous roots and leaves, and these very strong and fragrant of the odour of the mint, but the heat of my chamber, as I suppose, killed the plant when it was grown to have a pretty thick stalk, which with the various and ramified roots, which it shot into the water as if it had been earth, presented in its transparent flower-pot a spectacle not unpleasant to behold. The like I tried with sweet marjoram, and I found the experiment succeed also, though somewhat more slowly, with balm and peniroyal, to name now no other plants. And one of these vegetables, cherished only by water, having obtained a competent growth, I did, for tryals sake, cause to be distilled in a small retort, and thereby obtained some phlegme, a little empyreumaticall spirit, a small quantity of adult oyl, and a *caput mortuum*; which appearing to be a coal, I concluded it to consist of salt and earth: but the quantity of it was so small, that I forbore to calcine it. The water I used to nourish this plant was not shifted nor renewed; and I chose spring-water rather than rain-water, because the latter is more discernably a kind of πανσωερμία, which, though it be granted to be freed from grosser mixtures, seems yet to contain in it, besides the steams of several bodies wandering in the air, which may be supposed to impregnate it, a certain spirituous substance, which may be extracted out of it, and is by some mistaken for the spirit of the world corporifyed, upon what grounds, and with what probability, I may elsewhere perchance, but must not now, discourse to you.

But perhaps I might have saved a great part of my labour. For I finde that Helmont (an author more considerable for his experiments than many learned men are pleased to think him) having had an opportunity to prosecute an experiment much of the same nature with those I have been now speaking of, for five years together, obtained at the end of that time so notable a quantity of transmuted water, that I should scarce think it fit to have his experiment and mine mentioned together, were it not that the length of time requisite to this may deterr the curiosity of some, and exceed the leasure of others; and partly, that so paradoxical a truth as that which these experiments seem to hold forth, need to be confirmed by more witnesses than one, especially

since the extravagancies and untruths to be met with
in Helmont's treatise of the Magnetick Cure of Wounds,
have made his testimonies suspected in his other writings,
though as to some of the unlikely matters of fact he
delivers in them, I might safely undertake to be his
compurgator. But that experiment of his which I was
mentioning to you, he saies, was this. He took 200 pound
of earth dried in an oven, and having put it into an
earthen vessel and moistened it with rain water, he
planted in it the trunk of a willow tree of five pound
weight; this he watered, as need required, with rain or
with distilled water; and to keep the neighbouring earth
from getting into the vessel, he employed a plate of iron
tinned over and perforated with many holes. Five years
being effluxed, he took out the tree and weighed it, and
(with computing the leaves that fell during four autumnes)
he found it to weigh 169 pound, and about three ounces.
And having again dried the earth it grew in, he found it
want of its former weight of 200 pound, about a couple
only of ounces; so that 164 pound of the roots, wood,
and bark, which constituted the tree, seem to have sprung
from the water. And though it appears not that Helmont
had the curiosity to make any analysis of this plant,
yet what I lately told you I did to one of the vegetables
I nourished with water only, will I suppose keep you
from doubting that if he had distilled this tree, it would
have afforded him the like distinct substances as another
vegetable of the same kind. I need not subjoyne that
I had it also in my thoughts to try how experiments to the
same purpose with those I related to you would succeed
in other bodies than vegetables, because importunate
avocations having hitherto hindered me from putting my
design in practice, I can yet speak but conjecturally of
the success: but the best is, that the experiments already
made and mentioned to you need not the assistance of
new ones, to verifie as much as my present task makes it
concern me to prove by experiments of this nature.

One would suspect (saies Eleutherius after his long
silence) by what you have been discoursing, that you are
not far from Helmont's opinion about the origination of

compound bodies, and perhaps too dislike not the arguments which he imploys to prove it.

What Helmontian opinion, and what arguments do you mean (askes Carneades).

What you have been newly discoursing (replies Eleutherius) tells us, that you cannot but know that this bold and acute spagyrist scruples not to assert that all mixt bodies spring from one element; and that vegetables, animals, marchasites, stones, metalls, etc. are materially but simple water disguised into these various formes, by the plastick or formative vertue of their seeds. And as for his reasons you may find divers of them scattered up and down his writings; the considerablest of which seem to be these three; The ultimate reduction of mixt bodies into insipid water, the vicissitude of the supposed elements, and the production of perfectly mixt bodies out of simple water. And first he affirmes that the *sal circulatus Paracelsi*, or his liquor alkahest, does adequately resolve plants, animals, and mineralls into one liquor or more, according to their several internall disparities of parts, (without *caput mortuum*, or the destruction of their seminal vertues;) and that the alkahest being abstracted from these liquors in the same weight and vertue wherewith it dissolved them, the liquors may by frequent cohobations from chalke or some other idoneous matter, be totally deprived of their seminal endowments, and return at last to their first matter, insipid water; some other wayes he proposes here and there to divest some particular bodies of their borrowed shapes, and make them remigrate to their first simplicity. The second topick whence Helmont drawes his arguments, to prove water to be the material cause of mixt bodies, I told you was this, that the other supposed elements may be transmuted into one another. But the experiments by him here and there produced on this occasion, are so uneasie to be made and to be judged of, that I shall not insist on them; not to mention, that if they were granted to be true, his inference from them is somewhat disputable; and therefore I shall pass on to tell you, that as, in his first argument, our paradoxical author endeavours to prove water the sole element of mixt bodies, by their ultimate resolution, when by his alkahest, or some other conquering agent, the seeds have been destroyed, which disguised them; or when by time those seeds are wearied, or exantlated, or unable to act their parts upon the stage of the universe any longer: so in his third argument he endeavours to evince the same conclusion, by the constitution of bodies which he asserts to be nothing but water subdued by seminal vertues. Of this he gives here and there in his writings several instances, as to plants and animals; but divers of them being difficult either to be tried or to be understood, and others of them being not altogether unobnoxious to exceptions, I think you have singled out the principal and less questionable experiment when you lately mentioned, that of the willow tree. And having thus, continues Eleutherius, to answer your question, given you a summary account of what I am confident, you know better than I do, I shall be very glad to receive your sence of it, if the giving it me will not too much divert you from the prosecution of your discourse.

That *if* (replies Carneades) was not needlessly annexed: for thorowly to examine such an hypothesis and such arguments would require so many considerations, and consequently so much time, that I should not now have the leasure to perfect such a digression, and much less to finish my principal discourse. Yet thus much I shall tell you at present, that you need not fear my rejecting this opinion for its novelty; since, however the Helmontians may in complement to their master pretend it to be a new discovery, yet though the arguments be for the most part his, the opinion itself is very antient: for Diogenes Laertius and divers other authors speak of Thales, as the first among the Græcians that made disquisitions upon nature. And of this Thales, I remember, Tully informs us, that he taught all things were at first made of water. And it seems by Plutarch and Justin Martyr, that the opinion was ancienter than he: for they tell us that he used to defend his tenent by the testimony of Homer.

And a Greek author, the (Scholiast of Apollonius) upon these words

'Εξ ἰλύφ ἐβλάsησε χθὼν ἄυτη.

The earth of slime was made,

affirms, (out of Zeno) that the chaos, whereof all things were made, was, according to Hesiod, water; which, setling first, became slime, and then condensed into solid earth. And the same opinion about the generation of slime seems to have been entertained by Orpheus, out of whom one of the antients cites this testimony,

'Εχ τοῦ ὕδατφ ἰλὺs κατέsη.

Of water slime was made.

It seems also by what is delivered in Strabo out of another author concerning the Indians, that they likewise held that all things had differing beginnings, but that of which the world was made, was water. And the like opinion has been by some of the antients ascribed to the Phœnicians, from whom Thales himself is conceived to have borrowed it; as probably the Greeks did much of theologie, and, as I am apt to think, of their philosophy too; since the devising of the atomical hypothesis commonly ascribed to Leucippus and his disciple Democritus, is by learned men attributed to one Moschus a Phœnician. And possibly the opinion is yet antienter than so; for 'tis known that the Phœnicians borrowed most of their learning from the Hebrews. And among those that acknowledge the Books of Moses, many have been inclined to think water to have been the primitive and universal matter, by perusing the beginning of Genesis, where the waters seem to be mentioned as the material cause, not only of sublunary compound bodies, but of all those that make up the universe; whose component parts did orderly, as it were, emerge out of that vast abysse, by the operation of the Spirit of God, who is said to have been moving Himself, as hatching females do (as the original, *Merahephet*, is said to import, and it seems to signifie in one of the two other places, wherein alone I have met with it in the Hebrew Bible) upon the face of the waters; which being, as may be supposed, divinely impregnated with the seeds of all things, were by that productive incubation qualified to produce them. But you, I presume, expect that I should discourse of this matter like a naturalist, not a philologer. Wherefore I shall add, to countenance Helmont's opinion, that whereas he gives not, that I remember, any instance of any mineral body, nor scarce of any animal, generated of water, a French chymist, Monsieur de Rochas, has presented his readers an experiment, which if it were punctually such as he has delivered it, is very notable. He then, discoursing of the generation of things according to certain chymical and metaphorical notions (which I confess are not to me intelligible) sets down, among divers speculations not pertinent to our subject, the following narrative, which I shall repeat to you the sence of in English, with as little variation from the literal sence of the French words, as my memory will enable me. " Having (saies he) discerned such great wonders by the natural operation of water, I would know what may be done with it by art imitating nature. Wherefore I took water which I well knew not to be compounded, nor to be mixed with any other thing than that spirit of life (whereof he had spoken before) and with a heat artificial, continual and proportionate, I prepared and disposed it by the above-mentioned graduations of coagulation, congelation, and fixation, untill it was turned into earth, which earth produced animals, vegetables and minerals. I tell not what animals, vegetables and minerals, for that is reserved for another occasion: but the animals did move of themselves, eat, etc.—and by the true anatomie I made of them, I found that they were composed of much sulphur, little mercury, and less salt.—The minerals began to grow and increase by converting into their own nature one part of the earth thereunto disposed; they were solid and heavy. And by this truly demonstrative science, namely chymistry, I found that they were composed of much salt, little sulphur, and less mercury.

But (saies Carneades) I have some suspitions concerning this strange relation, which make me unwilling to declare an opinion of it, unless I were satisfied concerning divers material circumstances that our author has left unmentioned; though as for the generation of living creatures, both vegetable and sensitive, it needs not seem incredible, since we find that our common water (which indeed is often impregnated with variety of seminal principles and rudiments) being long kept in a quiet place will putrifie and stink, and then perhaps too produce moss and little worms, or other insects, according to the nature of the seeds that were lurking in it. I must likewise desire you to take notice, that as Helmont gives us no instance of the production of minerals out of water, so the main argument that he employs to prove that they and other bodies may be resolved into water, is drawn from the operations of his alkahest, and consequently cannot be satisfactorily examined by you and me.

Yet certainly (saies Eleutherius) you cannot but have somewhat wondered as well as I, to observe how great a share of water goes to the making up of divers bodies, whose disguises promise nothing neer so much. The distillation of eeles, though it yielded me some oyle, and spirit, and volatile salt, besides the *caput mortuum,* yet were all these so disproportionate to the phlegm that came from them, (and in which at first they boyled as in a pot of water) that they seemed to have bin nothing but coagulated phlegm, which does likewise strangely abound in vipers, though they are esteemed very hot in operation, and will in a convenient air survive some dayes the loss of their heads and hearts, so vigorous is their vivacity. Mans bloud itself as spirituous, and as elaborate a liquor as 'tis reputed, does so abound in phlegm, that, the other day, distilling some of it on purpose to try the experiment (as I had formerly done in deers bloud) out of about seven ounces and a halfe of pure bloud we drew neere six ounces of phlegm, before any of the more operative principles began to arise and invite us to change the receiver. And to satisfie myself that some of these animall phlegms were void enough of spirit to deserve that name, I would not content myself to taste them only, but fruitlessly poured on them acid liquors, to try if they contained any volatile salt or spirit, which (had there been any there) would probably have discovered itself by making an ebullition with the affused liquor. And now I mention corrosive spirits, I am minded to inform you, that though they seem to be nothing else but fluid salts, yet they abound in water, as you may observe, if either you entangle, and so fix their saline part, by making them corrode some idoneous body, or else if you mortifie it with a contrary salt; as I have very manifestly observed in the making a medicine somewhat like Helmont's *balsamus samech,* with distilled vinegar instead of spirit of wine, wherewith he prepares it: for you would scarce believe (what I have lately observed) that of that acid spirit, the salt of tartar, from which it is distilled, will by mortifying and retaining the acid salt turn into worthless phlegm neere twenty times its weight; before it be so fully impregnated as to rob no more distilled vinegar of its salt. And though spirit of wine exquisitely rectified seem of all liquors to be the most free from water, it being so igneous that it will flame all away without leaving the least drop behinde it, yet even this fiery liquor is by Helmont not improbably affirmed, in case what he relates be true, to be materially water, under a sulphureous disguise: for, according to him, in the making that excellent medicine Paracelsus his *balsamus samech,* (which is nothing but *sal tartari* dulcified by distilling from it spirit of wine till the salt be sufficiently glutted with its sulphur, and till it suffer the liquor to be drawn off, as strong as it was poured on) when the salt of tartar from which it is distilled hath retained, or deprived it of the sulphureous parts of the spirit of wine, the rest, which is incomparably the greater part of the liquor, will remigrate into phlegm. I added that clause [*in case what he relates be true*] because I have not as yet sufficiently tried it myself. But not only something of experiment keeps me from thinking it, as many chymists do, absurd, (though I have as well as they, in vain tried it with ordinary salt of tartar) but besides that Helmont often relates it, and draws consequences from it; a person noted for his soberness and skill in spagyrical preparations, having been askt by me whether the experiment might not be made to succeed, if the salt and spirit were prepared according to a way suitable to my principles, he affirmed to me, that he had that way I proposed made Helmont's experiment succeed very well, without adding anything to the salt and spirit. But our way is neither short nor easie.

I have indeed (saies Carneades) sometimes wondered to see how much phlegme may be obtained from bodies by the fire. But concerning that phlegme I may anon have occasion to note something, which I therefore shall not now anticipate. But to return to the opinion of Thales, and of Helmont, I consider, that supposing the alkahest could reduce all bodies into water, yet whether that water, because insipid, must be elementary, may not groundlesly be doubted; for I remember the candid and eloquent Petrus Laurembergius, in his notes upon Sala's aphorismes, affirmes that he saw an insipid menstruum that was a powerfull dissolvent, and (if my memory does not much mis-inform me) could dissolve gold. And the water which may be drawn from quicksilver without addition, though it be almost tasteless, you will I believe think of a differing nature from simple water, especially if you digest in it appropriated mineralls. To which I shall add but this, that this consideration may be further extended. For I see no necessity to conceive that the water mentioned in the beginning of Genesis, as the universal matter, was simple and elementary water; since though we should suppose it to have been an agitated congeries or heap consisting of a great variety of seminal principles and rudiments, and of other corpuscles fit to be subdued and fashioned by them, it might yet be a body fluid like water, in case the corpuscles it was made up of, were by their creator made small enough, and put into such an actuall motion as might make them glide along one another. And as we now say, the sea consists of water, (notwithstanding the saline, terrestrial, and other bodies mingled with it,) such a liquor may well enough be called water, because that was the greatest of the known bodies whereunto it was like; though, that a body may be fluid enough to appear a liquor, and yet contain corpuscles of a very differing nature, you will easily believe, if you but expose a good qantity of vitriol in a strong vessel to a competent fire. For although it contains both aqueous, earthy, saline, sulphureous, and metalline corpuscles, yet the whole mass will at first be fluid like water, and boyle like a seething pot.

I might easily (continues Carneades) enlarge myself on such considerations, if I were now obliged to give you my judgment of the Thalesian, and Helmontian hypothesis. But whether or no we conclude that all things were at first generated of water, I may deduce from what I have tried concerning the growth of vegetables, nourished with water, all that I now proposed to myself or need at present to prove, namely that salt, spirit, earth, and even oyl (though that be thought of all bodies the most opposite to water) may be produced out of water; and consequently that a chymical principle as well as a peripatetick element, may (in some cases) be generated anew, or obtained from such a parcel of matter as was not endowed with the form of such a principle or element before.

And having thus, Eleutherius, evinced that 'tis possible that such substances as those that chymists are wont to call their *tria prima,* may be generated, anew: I must next endeavour to make it probable, that the operation of the fire does actually (sometimes) not only divide compounded bodies into small parts, but compound those parts after a new manner, whence consequently, for ought we know, there may emerge as well saline and sulphureous substances, as bodies of other textures. And perhaps it will assist us in our enquiry after the effects of the operations of the fire upon other bodies, to consider a little, what it does to those mixtures which being productions of the art of man, we best know the composition of. You may then be pleased to take notice that though sope is made up by the sope-boylers of oyle or grease, and salt, and water diligently incorporated together; yet if you expose the mass they constitute to a graduall fire in a retort, you shall then indeed make a separation, but not

of the same substances that were united into sope, but of
others of a distant and yet not an elementary nature, and
especially of an oyle very sharp and fœtid, and of a very
differing quality from that which was employed to make
the sope: so, if you mingle in a due proportion, sal
armoniack with quick-lime, and distill them by degrees
of fire, you shall not divide the sal armoniack from the
quick-lime, though the one be a volatile, and the other
a fixed substance, but that which will ascend will be a
spirit much more fugitive, penetrant, and stinking, than
sal armoniack; and there will remain with the quick-lime
all, or very near all the sea salt, that concurred to make
up the sal armoniack; concerning which sea salt I shall,
to satisfie you how well it was united to the lime, informe
you, that I have by making the fire at length very vehement, caused both the ingredients to melt in the retort
itself into one mass, and such masses are apt to relent in
the moist air.   If it be here objected, that these instances
are taken from factitious concretes which are more
compounded than those which nature produces; I shall
reply, that besides that I have mentioned them as much
to illustrate what I proposed, as to prove it; it will be
difficult to evince that nature herself does not make
decompounded bodies, I mean, mingle together such mixt
bodies, as are already compounded of elementary, or
rather of more simple ones.   For vitriol (for instance)
though I have sometimes taken it out of minerall earths,
where nature had without any assistance of art prepared
it to my hand, is really, though chymists are pleased to
reckon it among salts, a decompounded body consisting
(as I shall have occasion to declare anon) of a terrestriall
substance, of a metal, and also of at least one saline body,
of a peculiar, and not elementary nature.   And we see
also in animals, that their blood may be composed of
divers very differing mixt bodies, since we find it observed
that divers sea-fowle taste rank of the fish on which they
ordinarily feed; and Hippocrates himself observes, that
a child may be purged by the milke of the nurse, if she
have taken elaterium; which argues that the purging
corpuscles of the medicament concurr to make up the

milk of the nurse; and that white liquor is generally by physitians supposed to be but blanched and altered blood. And I remember I have observed, not farr from the Alps, that at a certain time of the yeare the butter of that country was very offensive to strangers, by reason of the rank taste of a certain herb, whereon the cows were then wont plentifully to feed. But (proceeds Carneades) to give you instances of another kind, to shew that things may be obtained by the fire from a mixt body that were not pre-existent in it, let me remind you, that from many vegetables there may without any addition be obtained glass, a body, which I presume you will not say was pre-existent in it, but produced by the fire. To which I shall add but this one example more, namely that by a certain artificial way of handling quicksilver, you may without addition separate from it at least a 5th or 4th part of clear liquor, which with an ordinary peripatetick would pass for water, and which a vulgar chymist would not scruple to call phlegme, and which, for ought I have yet seen or heard, is not reducible into mercury again, and consequently is more than a disguise of it. Now besides that divers chymists will not allow mercury to have any, or at least any considerable quantity of either of the ignoble ingredients, earth and water; besides this, I say, the great ponderousness of quicksilver makes it very unlikely that it can have so much water in it as may be thus obtained from it, since mercury weighs 12 or 14 times as much as water of the same bulk. Nay for a further confirmation of this argument, I will add this strange relation, that two friends of mine, the one a physitian, and the other a mathematician, and both of them persons of unsuspected credit, have solemnly assured me, that after many tryals they made, to reduce mercury into water, in order to a philosophicall work, upon gold (which yet, by the way, I know proved unsuccessfull) they did once by divers cohobations reduce a pound of quicksilver into almost a pound of water, and this without the addition of any other substance, but only by pressing the mercury by a skilfully managed fire in purposely contrived vessels. But of these experiments our friend (saies Carneades, pointing at the register of this dialogue) will perhaps give you a more particular account than it is necessary for me to do: since what I have now said may sufficiently evince, that the fire may sometimes as well alter bodies as divide them, and by it we may obtain from a mixt body what was not pre-existent in it. And how are we sure, that in no other body what we call phlegme is barely separated, not produced by the action of the fire: since so many other mixt bodies are of a much less constant, and more alterable nature, than mercury (by many tricks it is wont to put upon chymists, and by the experiments I told you of, about an hour since) appears to be. But because I shall ere long have occasion to resume into consideration the power of the fire to produce new concretes, I shall no longer insist on this argument at present; only I must mind you, that if you will not disbelieve Helmont's relations, you must confess that the *tria prima* are neither ingenerable nor incorruptible substances; since by his alkahest some of them may be produced of bodies that were before of another denomination; and by the same powerfull menstruum all of them may be reduced into insipid water.

Here Carneades was about to pass on to his third consideration, when Eleutherius being desirous to hear what he could say to clear his second general consideration from being repugnant to what he seemed to think the true theory of mistion, prevented him by telling him, I somewhat wonder, Carneades, that you, who are in so many points unsatisfied with the peripatetick opinion touching the elements and mixt bodies, should also seem averse to that notion touching the manner of mistion, wherein the chymists (though perhaps without knowing that they do so) agree with most of the antient philosophers that preceded Aristotle, and that for reasons so considerable, that divers modern naturalists and physitians, in other things unfavourable enough to the spagyrists, do in this case side with them against the common opinion of the schools. If you should ask me (continues Eleutherius) what reasons I mean? I should partly by the writings of Sennertus and other learned men, and partly by my own thoughts, be supplied with more, than 'twere at present proper for me to insist largely on. And therefore I shall mention only, and that briefly, three or four. Of these, I shall take the first from the state of the controversie itself, and the genuine notion of mistion, which though much intricated by the schoolmen, I take in short to be this. Aristotle, at least as many of his interpreters expound him, and as indeed he teaches in some places, where he professedly dissents from the antients, declares mistion to be such a mutual penetration, and perfect union of the mingled elements, that there is no portion of the mixt body, how minute soever, which does not contain all, and every of the four elements, or in which, if you please, all the elements are not. And I remember, that he reprehends the mistion taught by the ancients, as too slight or gross, for this reason, that bodies mixt according to their hypothesis, though they appear to humane eyes, would not appear such to the acute eyes of a lynx, whose perfecter sight would discerne the elements, if they were no otherwise mingled, than as his predecessors would have it, to be but blended, not united; whereas the antients, though they did not all agree about what kind of bodies were mixt, yet they did almost unanimously hold, that in a compounded bodie, though the *miscibilia*, whether elements, principles, or whatever they pleased to call them, were associated in such small parts, and with so much exactness, that there was no sensible part of the mass but seemed to be of the same nature with the rest, and with the whole; yet as to the atomes, or other insensible parcels of matter, whereof each of the *miscibilia* consisted, they retained each of them its own nature, being but by apposition or juxta-position united with the rest into one bodie. So that although by vertue of this composition the mixt body did perhaps obtain divers new qualities, yet still the ingredients that compounded it, retaining their own nature, were by the destruction of the *compositum* separable from each other, the minute parts disingaged from those of a differing nature, and associated with those of their own sort returning to be again, fire, earth,

or water, as they were before they chanced to be ingredients of that *compositum.* This may be explained
(continues Eleutherius) by a piece of cloath made of white
and black threds interwoven, wherein though the whole
piece appear neither white nor black, but of a resulting
colour, that is gray, yet each of the white and black threds
that compose it, remains what it was before, as would
appear if the threds were pulled asunder, and sorted each
colour by itself. This (pursues Eleutherius) being, as I
understand it, the state of the controversie, and the
Aristotelians after their master commonly defining, that
mistion is *miscibilium alteratorum unio,* that seems to
comport much better with the opinion of the chymists,
than with that of their adversaries, since according to
that as the newly mentioned example declares, there is
but a juxta-position of separable corpuscles, retaining
each its own nature, whereas according to the Aristotelians,
when what they are pleased to call a mixt body results
from the concourse of the elements, the *miscibilia* cannot
so properly be said to be altered, as destroyed, since there
is no part in the mixt body, how small soever, that can
be called either fire, or air, or water, or earth.

Nor indeed can I well understand, how bodies can be
mingled other waies than as I have declared, or at least
how they can be mingled, as our peripateticks would
have it. For whereas Aristotle tells us, that if a drop of
wine be put into ten thousand measures of water, the wine
being overpowered by so vast a quantity of water will
be turned into it, he speaks to my apprehension, very
improbably. For though one should add to that quantity
of water as many drops of wine as would a thousand times
exceed it all, yet by his rule the whole liquor should not
be a *crama,* a mixture of wine and water, wherein the
wine would be predominant, but water only; since the
wine being added but by a drop at a time, would still fall
into nothing but water, and consequently would be turned
into it. And if this would hold in metals too, 'twere a
rare secret for goldsmiths, and refiners; for by melting
a mass of gold, or silver, and by but casting into it lead
or antimony, grain after grain, they might at pleasure,

within a reasonable compass of time, turn what quantity they desire, of the ignoble into the noble metalls. And indeed since a pint of wine, and a pint of water, amount to about a quart of liquor, it seems manifest to sense, that these bodies doe not totally penetrate one another, as one would have it; but that each retains its own dimensions; and consequently, that they are by being mingled only divided into minute bodies, that do but touch one another with their surfaces, as do the grains of wheat, rye, barley, etc. in a heap of severall sorts of corn: and unless we say, that as when one measure of wheat, for instance, is blended with a hundred measures of barely, there happens only a juxta-position and superficial contact betwixt the grains of wheat, and as many or thereabouts of the grains of barley; so when a drop of wine is mingled with a great deal of water, there is but an apposition of so many vinous corpuscles to a correspondent number of aqueous ones; unless I say this be said, I see not how that absurdity will be avoyded, whereunto the Stoical notion of mistion (namely by σύγχυσις, or confusion) was liable, according to which the least body may be co-extended with the greatest: since in a mixt body wherein before the elements were mingled there was, for instance, but one pound of water to ten thousand of earth, yet according to them there must not be the least part of that compound, that consisted not as well of earth, as water. But I insist, perhaps, too long (saies Eleutherius) upon the proofs afforded me by the nature of mistion: wherefore I will but name two or three other arguments; whereof the first shall be, that according to Aristotle himself, the motion of a mixt body followes the nature of the predominant element, as those wherein the earth prevails, tend towards the centre of heavy bodies. And since many things make it evident, that in divers mixt bodies the elementary qualities are as well active, though not altogether so much so as in the elements themselves, it seems not reasonable to deny the actual existence of the elements in those bodies wherein they operate.

To which I shall add this convincing argument, that experience manifests, and Aristotle confesses it, that the *miscibilia* may be again separated from a mixt body, as is obvious in the chymical resolutions of plants and animalls, which could not be unless they did actually retain their formes in it: for since, according to Aristotle, and I think according to truth, there is but one common mass of all things, which he has been pleased to call *materia prima ;* and since 'tis not therefore the matter but the forme that constitutes and discriminates things, to say that the elements remain not in a mixt body, according to their formes, but according to their matter, is not to say that they remain there at all; since although those portions of matter were earth and water, etc. before they concurred; yet the resulting body being once constituted, may as well be said to be simple as any of the elements; the matter being confessedly of the same nature in all bodies, and the elementary formes being according to this hypothesis perished and abolished.

And lastly, and if we will consult chymical experiments, we shall find the advantages of the chymical doctrine above the peripatetick title little less than palpable. For in that operation that refiners call quartation, which they employ to purifie gold, although three parts of silver be so exquisitely mingled by fusion with a fourth part of gold (whence the operation is denominated) that the resulting mass acquires several new qualities, by vertue of the composition, and that there is scarce any sensible part of it that is not composed of both the metalls; yet if you cast this mixture into *aqua fortis,* the silver will be dissolved in the menstruum, and the gold like a dark or black powder will fall to the bottom of it, and either body may be again reduced into such a metal as it was before; which shews, that it retained its nature, notwithstanding its being mixt *per minima* with the other: we likewise see, that though one part of pure silver be mingled with eight or ten parts, or more, of lead; yet the fire will upon the cuppel easily and perfectly separate them again. And that which I would have you peculiarly consider on this occasion is, that not only in chymicall anatomies there is a separation made of the elementary ingredients, but that some mixt bodies afford a very much greater quantity of this or that element or principle, than of another; as we see, that turpentine and amber yeeld much more oyl and sulphur than they do water; whereas wine, which is confessed to be a perfectly mixt bodie, yeelds but a little inflamable spirit, or sulphur, and not much more earth; but affords a vast proportion of phlegm or water: which could not be, if, as the peripateticks suppose, every, even of the minutest particles, were of the same nature with the whole, and consequently did contain both earth and water, and aire, and fire; wherefore as to what Aristotle principally, and almost only objects, that unless his opinion be admitted, there would be no true and perfect mistion, but onely aggregates or heaps of contiguous corpuscles, which, though the eye of man cannot discerne, yet the eye of a lynx might perceive not to be of the same nature with one another and with their *totum*, as the nature of mistion requires, if he do not beg the question, and make mistion to consist in what other naturalists deny to be requisite to it, yet he at least objects that as a great inconvenience which I cannot take for such, till he have brought as considerable arguments as I have proposed to prove the contrary, to evince that nature makes other mistions than such as I have allowed, wherein the *miscibilia* are reduced into minute parts, and united as far as sense can discerne: which if you will not grant to be sufficient for a true mistion, he must have the same quarrel with nature herself, as with his adversaries.

Wherefore (continues Eleutherius) I cannot but somewhat marvail that Carneades should oppose the doctrine of the chymists in a particular, wherein they do as well agree with his old mistress, nature, as dissent from his old adversary, Aristotle.

I must not (replies Carneades) engage myself at present to examine throughly the controversies concerning mistion: and if there were no third thing, but that I were reduced to embrace absolutely and unreservedly either the opinion of Aristotle, or that of the philosophers that went before him, I should look upon the latter, which the chymists have adopted, as the more defensible opinion: but because differing in the opinions about the elements from both parties, I think I can take a middle course, and discourse to you of mistion after a way that does neither perfectly agree, nor perfectly disagree with either, as I will not peremtorily define, whether there be not cases wherein some phænomena of mistion seem to favour the opinion that the chymists patrons borrowed of the antients, I shall only endeavour to shew you that there are some cases which may keep the doubt, which makes up my second general consideration from being unreasonable.

I shall then freely acknowledge to you (saies Carneades) that I am not over-well satisfied with the doctrine that is ascribed to Aristotle, concerning mistion, especially since it teaches that the four elements may again be separated from the mixt body; whereas if they continued not in it, it would not be so much a separation as a production. And I think the ancient philosophers that preceded Aristotle, and chymists who have since received the same opinion, do speak of this matter more intelligibly, if not more probably, than the peripateticks: but though they speak congruously enough, to their believing, that there are a certain number of primogeneal bodies, by whose concourse all those we call mixt are generated, and which in the destruction of mixt bodies do barely part company, and reduce from one another, just such as they were when they came together; yet I, who meet with very few opinions that I can entirely acquiesce in, must confess to you that I am inclined to differ not only from the Aristotelians, but from the old philosophers and the chymists, about the nature of mistion: and if you will give me leave, I shall briefly propose to you my present notion of it, provided you will look upon it, not so much as an assertion as an hypothesis; in talking of which I do not now pretend to propose and debate the whole doctrine of mistion, but to shew that 'tis not improbable, that sometimes mingled substances may be so strictly united, that it doth not by the usuall operations of the fire, by which chymists are wont to suppose themselves to have made the analysis of mixt bodies, sufficiently appear, that in such bodies the *miscibilia*, that concurred to make them up, do each of them retain its own peculiar nature; and by the spagyrists fires may be more easily extricated and recovered, than altered, either by a change of texture in the parts of the same ingredient, or by an association with some parts of another ingredient more strict than was that of the parts of this or that *miscibile* among themselves. At these words Eleu. having pressed him to do what he proposed, and promised to do what he desired;

I consider then (resumes Carneades) that, not to mention those improper kinds of mistion, wherein *homogeneous* bodies are joyned, as when water is mingled with water, or two vessels full of the same kind of wine with one another, the mistion I am now to discourse of seems, generally speaking, to be but an union *per minima* of any two or more bodies of differing denominations; as when ashes and sand are colliquated into glass; or antimony and iron into *regulus martis ;* or wine and water are mingled, and sugar is dissolved in the mixture. Now in this general notion of mistion it does not appear clearly comprehended, that the *miscibilia* or ingredients do in their small parts so retain their nature and remain distinct in the compound, that they may thence by the fire be again taken asunder: for though I deny not that in some mistions of certain permanent bodies this recovery of the same ingredients may be made; yet I am not convinced that it will hold in all or even in most, or that it is necessarily deducible from chymicall experiments, and the true notion of mistion. To explain this a little, I assume, that bodies may be mingled, and that very durably, that are not elementary, nor have been resolved into elements or principles, that they may be mingled; as is evident in the *regulus* of colliquated antimony, and iron newly mentioned; and in gold coyne, which lasts so many ages; wherein generally the gold is alloyed by the mixture of a quantity, greater or lesser, (in our mints they use about a 12th part) of either silver, or copper, or both. Next, I consider, that there being but one universal matter of things, as 'tis known that the Aristotelians themselves acknowledge, who call it *materia prima* (about which

nevertheless I like not all their opinions) the portions of
this matter seem to differ from one another, but in certain
qualities or accidents, fewer or more; upon whose account
the corporeal substance they belong to receives its denomination, and is referred to this or that particular sort of
bodies; so that if it come to lose, or be deprived of those
qualities, though it ceases not to be a body, yet it ceases
from being that kind of body as a plant, or animal, or
red, green, sweet, sowre, or the like. I consider that it
very often happens that the small parts of bodies cohere
together but by immediate contact and rest, and that
however, there are few bodies whose minute parts stick
so close together, to what cause soever their combination
be ascribed, but that it is possible to meet with some other
body, whose small parts may get between them, and so
disjoyn them; or may be fitted to cohere more strongly
with some of them, than those some do with the rest; or
at least may be combined so closely with them, as that
neither the fire, nor the other usual instruments of
chymical anatomies will separate them. These things
being premised, I will not peremptorily deny, but that
there may be some clusters of particles, wherein the
particles are so minute, and the coherence so strict, or
both, that when bodies of differing denominations, and
consisting of such durable clusters, happen to be mingled,
though the compound body made up of them may be very
differing from either of the ingredients, yet each of the
little masses or clusters may so retain its own nature, as
to be again separable, such as it was before. As when
gold and silver being melted together in a due proportion
(for in every proportion, the refiners will tell you that the
experiment will not succeed) *aqua fortis* will dissolve the
silver, and leave the gold untoucht; by which means, as
you lately noted, both the metalls may be recovered from
the mixed mass. But (continues Carneades) there are other
clusters wherein the particles stick not so close together, but
that they may meet with corpuscles of another denomination, which are disposed to be more closely united with some
of them, than they were among themselves. And in such
case, two thus combining corpuscles losing that shape, or

size, or motion, or other accident, upon whose account
they were endowed with such a determinate quality or
nature, each of them really ceases to be a corpuscle of the
same denomination it was before; and from the coalition
of these there may emerge a new body, as really one, as
either of the corpuscles was before they were mingled, or,
if you please, confounded: since this concretion is really
endowed with its own distinct qualities, and can no more
by the fire, or any other known way of analysis, be
divided again into the corpuscles that at first concurred
to make it, than either of them could by the same means
be subdivided into other particles.　But (saies Eleutherius)
to make this more intelligible by particular examples;
If you dissolve copper in *aqua fortis*, or spirit of nitre, (for
I remember not which I used, nor do I think it much
material) you may by chrystalising the solution obtain
a goodly vitriol; which though by vertue of the composition it have manifestly diverse qualities, not to be
met with in either of the ingredients, yet it seems that
the nitrous spirits, or at least many of them, may in this
compounded mass retain their former nature; for having
for tryal sake distilled this vitriol spirit, there came over
store of red fumes, which by that colour, by their peculiar
stinke, and by their sowrness, manifested themselves to
be, nitrous spirits; and that the remaining calx continued
copper, I suppose you'll easily believe.　But if you
dissolve minium, which is but lead powdered by the fire,
in good spirit of vinegar, and chrystalise the solution,
you shall not only have a saccharine salt exceedingly
differing from both its ingredients; but the union of some
parts of the menstruum with some of those of the metal
is so strict, that the spirit of vinegar seems to be, as such,
destroyed; since the saline corpuscles have quite lost
that acidity, upon whose account the liquor was called
spirit of vinegar; nor can any such acid parts as were
put to the minium be separated by any known way from
the *saccharum saturni* resulting from them both; for not
only there is no sowrness at all, but an admirable sweetness
to be tasted in the concretion; and not only I found not
that spirit of wine, which otherwise will immediately hiss

when mingled with strong spirit of vinegar, would hiss being poured upon *saccharum saturni*, wherein yet the acid salt of vinegar, did it survive, may seem to be concentrated; but upon the distillation of *saccharum saturni* by itself I found indeed a liquor very penetrant, but not at all acid, and differing as well in smell and other qualities, as in taste, from the spirit of vinegar; which likewise seemed to have left some of its parts very firmly united to the *caput mortuum*, which though of a leaden nature was in smell, colour, etc. differing from minium; which brings into my mind, that though two powders, the one blew, and the other yellow, may appear a green mixture, without either of them losing its own colour, as a good microscope has sometimes informed me; yet having mingled minium and sal armoniack in a requisite proportion, and exposed them in a glass vessel to the fire, the whole mass became white, and the red corpuscles were destroyed; for though the calcined lead was separable from the salt, yet you'll easily believe it did not part from it in the forme of a red powder, such as was the minium, when it was put to the sal armoniack. I leave it also to be considered, whether in blood, and divers other bodies, it be probable, that each of the corpuscles that concur to make a compound body doth, though some of them in some cases may, retain its own nature in it, so that chymists may extricate each sort of them from all the others, wherewith it concurred to make a body of one denomination.

I know there may be a distinction betwixt matter *immanent*, when the material parts remain and retain their own nature in the things materiated, as some of the schoolmen speak (in which sence wood, stones and lime are the matter of a house) and *transient*, which in the materiated thing is so altered, as to receive a new forme, without being capable of re-admitting again the old. In which sence the friends of this distinction say, that chyle is the matter of blood, and blood that of a humane body, of all whose parts 'tis presumed to be the aliment. I know also that it may be said, that of material principles, some are *common* to all mixt bodies, as Aristotle's four elements, or the chymists *tria prima*; others *peculiar*, which belong to this or that sort of bodies; as butter and a kind of whey may be said to be the proper principles of cream: and I deny not, but that these distinctions may in some cases be of use; but partly by what I have said already, and partly by what I am to say, you may easily enough guess in what sence I admit them, and discerne that in such a sence they will either illustrate some of my opinions, or at least will not overthrow any of them.

To prosecute then what I was saying before, I will add to this purpose, that since the major part of chymists credit, what those they call philosophers affirme of their stone, I may represent to them, that though when common gold and lead are mingled together, the lead may be severed almost unaltered from the gold; yet if instead of gold a tantillum of the red elixir be mingled with the saturn, their union will be so indissoluble in the perfect gold that will be produced by it, that there is no known, nor perhaps no possible way of separating the diffused elixir from the fixed lead, but they both constitute a most permanent body, wherein the saturn seems to have quite lost its properties that made it be called lead, and to have been rather transmuted by the elixir, than barely associated to it. So that it seems not alwaies necessary, that the bodies that are put together *per minima* should each retain its own nature; so as when the mass itself is dissipated by the fire, to be more disposed to re-appear in its pristine forme, than in any new one, which by a stricter association of its parts with those of some of the other ingredients of the *compositum*, than with one another, it may have acquired.

And if it be objected, that unless the hypothesis I oppose be admitted, in such cases as I have proposed, there would not be an union, but a destruction of mingled bodies, which seems all one as to say, that of such bodies there is no mistion at all; I answer, that though the substances that are mingled remain, only their accidents are destroyed, and though we may with tolerable congruity call them *miscibilia*, because they are distinct bodies before they are put together, however afterwards they are so confounded that I should rather call them concretions, or resulting bodies, than mixt ones; and though perhaps some other and better account may be proposed, upon which the name of mistion may remain; yet if what I have said be thought reason, I shall not wrangle about words, though I think it fitter to alter a terme of art, than reject a new truth, because it suits not with it. If it be also objected that this notion of mine, concerning mistion, though it may be allowed, when bodies already compounded are put to be mingled, yet it is not applicable to those mistions that are immediately made of the elements, or principles themselves; I answer in the first place, that I here consider the nature of mistion somewhat more generally, than the chymists; who yet cannot deny that there are oftentimes mixtures, and those very durable ones, made of bodies that are not elementary. And in the next place, that though it may be probably pretended that in those mixtures that are made immediately of the bodies, that are called principles or elements, the mingled ingredients may better retain their own nature in the compounded mass, and be more easily separated from thence; yet, besides that it may be doubted, whether there be any such primary bodies, I see not why the reason I alledged, of the destructibility of the ingredients of bodies in general, may not sometimes be applicable to salt, sulphur, or mercury; 'till it be shewn upon what account we are to believe them priviledged. And however, (if you please but to recall to mind, to what purpose I told you at first, I meant to speak of mistion at this time) you will perhaps allow, that what I have hitherto discoursed about it, may not only give some light to the nature of it in general (especially when I shall have an opportunity to declare to you my thoughts on that subject more fully) but may on some occasions also be serviceable to me in the insuing part of this discourse.

But to look back now to that part of our discourse, whence this excursion concerning mistion has so long diverted us, though we there deduced from the differing substances obtained from a plant nourished only with water, and from some other things, that it was not

necessary that nature should alwaies compound a body
at first of all such differing bodies as the fire could afterwards make it afford; yet this is not all that may be
collected from those experiments. For from them there
seems also deducible something that subverts another
foundation of the chymical doctrine. For since that (as
we have seen) out of fair water alone, not only spirit, but
oyle, and salt, and earth may be produced; it will follow
that salt and sulphur are not primogeneal bodies, and
principles, since they are every day made out of plain
water by the texture which the seed or seminal principle
of plants put it into. And this would not perhaps seem
so strange, if through pride or negligence, we were not
wont to overlook the obvious and familiar workings of
nature; for if we consider what slight qualities they are
that serve to denominate one of the *tria prima*, we shall
find that nature does frequently enough work as great
alterations in divers parcells of matter: for to be readily
dissoluble in water, is enough to make the body that is so,
pass for a salt. And yet I see not why from a new shufling
and disposition of the component particles of a body, it
should be much harder for nature to compose a body
dissoluble in water of a portion of water that was not so
before, than of the liquid substance of an egg, which will
easily mix with water, to produce by the bare warmth of
a hatching hen, membrans, feathers, tendons, and other
parts, that are not dissoluble in water as that liquid
substance was: nor is the hardness and brittleness of
salt more difficult for nature to introduce into such a
yielding body as water, than it is for her to make the
bones of a chick out of the tender substance of the liquors
of an egg. But instead of prosecuting this consideration,
as I easily might, I will proceed, as soon as I have taken
notice of an objection that lies in my way. For I easily
foresee it will be alledged, that the above mentioned
examples are all taken from plants, and animals, in whom
the matter is fashioned by the plastick power of the seed,
or something analogous thereunto. Whereas the fire
does not act like any of the seminal principles, but de-
stroyes them all when they come within its reach. But to

this I shall need at present to make but this easy answer, that whether it be a seminal principle, or any other which fashions that matter after those various manners I have mentioned to you, yet 'tis evident, that either by the plastick principle alone, or that and heat together, or by some other cause capable to contex the matter, it is yet possible that the matter may be anew contrived into such bodies. And 'tis only for the possibility of this that I am now contending.

THE THIRD PART

WHAT I have hitherto discoursed, Eleutherius (saies his
friend to him) has, I presume, shewn you, that a considering man may very well question the truth of those very
suppositions which chymists as well as peripateticks,
without proving, take for granted; and upon which
depends the validity of the inferences they draw from
their experiments. Wherefore having dispatched that,
which though a chymist perhaps will not, yet I do, look
upon as the most important, as well as difficult, part of my
task, it will now be seasonable for me to proceed to the
consideration of the experiments themselves, wherein
they are wont so much to triumph and glory. And these
will the rather deserve a serious examination, because
those that alledge them are wont to do it with so much
confidence and ostentation, that they have hitherto
imposed upon almost all persons, without excepting
philosophers and physitians themselves, who have read
their books, or heard them talk. For some learned men
have been content rather to believe what they so boldly
affirme, than be at the trouble and charge, to try whether
or no it be true. Others again, who have curiosity enough
to examine the truth of what is averred, want skill and
opportunity to do what they desire. And the generality
even of learned men, seeing the chymists (not contenting
themselves with the schools to amuse the world with empty
words) actually perform divers strange things, and,
among those resolve compound bodies into several substances not known by former philosophers to be contained
in them: men I say, seeing these things, and hearing
with what confidence chymists averr the substances
obtained from compound bodies by the fire to be the true
elements, or (as they speak) hypostatical principles of
them, are forward to think it but just as well as modest,
that according to the logicians rule, the skilfull artists

94

should be credited in their own art; especially when those things whose nature they so confidently take upon them to teach others, are not only productions of their own skill, but such as others know not else what to make of.

But though (continues Carneades) the chymists have been able upon some or other of the mentioned accounts, not only to delight but amaze, and almost to bewitch even learned men; yet such as you and I, who are not unpractised in the trade, must not suffer ourselves to be imposed upon by hard names, or bold assertions; nor to be dazled by that light which should but assist us to discern things the more clearly.   It is one thing to be able to help nature to produce things, and another thing to understand well the nature of the things produced.   As we see, that many persons that can beget children, are for all that as ignorant of the number and nature of the parts, especially the internal ones, that constitute a child's body, as they that never were parents.   Nor do I doubt, but you'll excuse me, if as I thank the chymists for the things their analysis shews me, so I take the liberty to consider how many, and what they are, without being astonisht at them; as if, whosoever hath skill enough to shew men some new thing of his own making, had the right to make them believe whatsoever he pleases to tell them concerning it.

Wherefore I will now proceed to my third general con-sideration, which is, that it does not appear, that *three* is precisely and universally the number of the distinct substances or elements, whereinto mixt bodies are resoluble by the fire, I mean that 'tis not proved by chymists, that all the compound bodies, which are granted to be perfectly mixt, are upon their chymical analysis divisible each of them into just three distinct substances, neither more nor less, which are wont to be lookt upon as elementary, or may as well be reputed so as those that are so reputed. Which last clause I subjoyne, to prevent your objecting that some of the substances I may have occasion to mention by and by, are not perfectly homogeneous, nor consequently worthy of the name of principles.   For that which I am now to consider, is, into how many differing substances, that may plausibly pass for the elementary ingredients of a mixed body, it may be analysed by the fire; but whether each of these be uncompounded, I reserve to examine, when I shall come to the next general consideration; where I hope to evince, that the substances which the chymists not only allow, but assert to be the component principles of the body resolved into them, are not wont to be uncompounded.

Now there are two kinds of arguments (pursues Carneades) which may be brought to make my third proposition seem probable; one sort of them being of a more speculative nature, and the other drawn from experience. To begin then with the first of these.

But as Carneades was going to do as he had said, Eleutherius interrupted him, by saying with a somewhat smiling countenance;

If you have no mind I should think, that the proverb, " That good wits have bad memories," is rational and applicable to you, you must not forget now you are upon the speculative considerations, that may relate to the number of the elements; that yourself did not long since deliver and concede some propositions in favour of the chymical doctrine, which I may without disparagement to you think it uneasie, even for Carneades to answer.

I have not, replies he, forgot the concessions you mean; but I hope too, that you have not forgot neither with what cautions they were made, when I had not yet assumed the person I am now sustaining. But however, I shall to content you, so discourse of my third general consideration, as to let you see, that I am not unmindful of the things you would have me remember.

To talk then again according to such principles as I then made use of, I shall represent, that if it be granted rational to suppose, as I then did, that the elements consisted at first of certain small and primary coalitions of the minute particles of matter into corpuscles very numerous, and very like each other, it will not be absurd to conceive, that such primary clusters may be of far more sorts than three or five; and consequently, that we need not suppose, that in each of the compound bodies

we are treating of, there should be found just three sorts
of such primitive coalitions, as we are speaking of.

And if according to this notion we allow a considerable
number of differing elements, I may add, that it seems
very possible, that to the constitution of one sort of mixt
bodies two kinds of elementary ones may suffice (as I lately
exemplified to you, in that most durable concrete, glass),
another sort of mixts may be composed of three elements,
another of four, another of five, and another perhaps of
many more. So that according to this notion, there can
be no determinate number assigned, as that of the elements,
of all sorts of compound bodies whatsoever, it being very
probable that some concretes consist of fewer, some of
more elements. Nay, it does not seem impossible, according to these principles, but that there may be two sorts
of mixts, whereof the one may not have any of all the same
elements as the other consists of; as we oftentimes see
two words, whereof the one has not any one of the letters
to be met with in the other; or as we often meet with
diverse electuaries, in which no ingredient (except sugar)
is common to any two of them. I will not here debate
whether there may not be a multitude of these corpuscles,
which by reason of their being primary and simple, might
be called elementary, if several sorts of them should convene to compose any body, which are as yet free, and
neither as yet contexed and entangled with primary
corpuscles of other kinds, but remains liable to be subdued
and fashioned by seminal principles, or the like powerful
and transmuting agent, by whom they may be so connected among themselves, or with the parts of one of the
bodies, as to make the compound bodies, whose ingredients
they are, resoluble into more, or other elements than those
that chymists have hitherto taken notice of.

To all which I may add, that since it appears, by what
I observed to you of the permanency of gold and silver,
that even corpuscles that are not of an elementary but
compounded nature, may be of so durable a texture, as to
remain indissoluble in the ordinary analysis that chymists
make of bodies by the fire; 'tis not impossible but that,
though there were but three elements, yet there may be

a greater number of bodies, which the wonted waies of anatomy will not discover to be no elementary bodies.

But, (saies Carneades) having thus far, in compliance to you, talket conjecturally of the number of the elements, 'tis now time to consider, not of how many elements it is possible that nature may compound mixed bodies, but (at least as far as the ordinary experiments of chymists will informe us) of how many she doth make them up.

I say then, that it does not by these sufficiently appear to me, that there is any one determinate number of elements to be uniformly met with in all the several sorts of bodies allowed to be perfectly mixt.

And for the more distinct proof of this proposition, I shall in the first place represent, that there are divers bodies, which I could never see by fire divided into so many as three elementary substances. I would fain (as I said lately to Philoponus) see that fixt and noble metal we call gold separated into salt, sulphur and mercury: and if any man will submit to a competent forfeiture in case of failing, I shall willingly in case of prosperous success pay for both the materials and the charges of such an experiment. 'Tis not, that after what I have tried myself I dare peremptorily deny, that there may out of gold be extracted a certain substance, which I cannot hinder chymists from calling its tincture or sulphur; and which leaves the remaining body deprived of its wonted colour. Nor am I sure, that there cannot be drawn out of the same metal a real quick and running mercury. But for the salt of gold, I never could either see it, or be satisfied that there was ever such a thing separated, *in rerum natura*, by the relation of any credible eye witness. And for the several processes that promise that effect, the materials that must be wrought upon are somewhat too precious and costly to be wasted upon so groundless adventures, of which not only the success is doubtful, but the very possibility is not yet demonstrated. Yet that which most deterrs me from such tryalls, is not their chargeableness, but their unsatisfactorinesse, though they should succeed. For the extraction of this golden salt being in chymists processes prescribed to be effected by corrosive menstruums, or the intervention of other saline bodies, it will remain doubtfull to a wary person, whether the emergent salt be that of the gold itself; or of the saline bodies or spirits employed to prepare it; for that such disguises of metals do often impose upon artists, I am sure Eleutherius is not so much a stranger to chymistry as to ignore. I would likewise willingly see the three principles separated from the pure sort of virgin-sand, from osteocalla, from refined silver, from quicksilver, freed from its adventitious sulphur, from Venetian talck, which by long detention in an extreme *reverberium*, I could but divide into smaller particles, not the constituent principles; nay, which, when I caused it to be kept, I know not how long, in a glass-house fire, came out in the figure it's lumps had when put in, though altered to an almost amethystine colour; and from divers other bodies, which it were now unnecessary to enumerate. For though I dare not absolutely affirme it to be impossible to analyze these bodies into their *tria prima*; yet because neither my own experiments, nor any competent testimony hath hitherto either taught me how such an analysis may be made, or satisfied me, that it hath been so, I must take the liberty to refrain from believing it, till the chymists prove it, or give us intelligible and practicable processes to perform what they pretend. For whilst they affect that ænigmatical obscurity with which they are wont to puzzle the readers of their divulged processes concerning the analytical preparation of gold or mercury, they leave wary persons much unsatisfied whether or no the differing substances, they promise to produce, be truly the hypostatical principles, or only some intermixtures of the divided bodies with those employed to work upon them, as is evident in the seeming chrystalls of silver, and those of mercury; which though by some inconsiderately, supposed to be the salts of those metals, are plainly but mixtures of the metalline bodies, with the saline parts of *aqua fortis* or other corrosive liquors; as is evident by their being reducible into silver or quicksilver, as they were before.

I cannot but confess (saith Eleutherius) that though

chymists may upon probable grounds affirme themselves able to obtain their *tria prima*, from animals and vegetables, yet I have often wondred that they should so confidently pretend also to resolve all metalline and other mineral bodies into salt, sulphur, and mercury. For 'tis a saying almost proverbial, among those chymists themselves that are accounted philosophers; and our famous countryman Roger Bacon has particularly adopted it; that, *facilius est aurum facere, quam destruere*. And I fear, with you, that gold is not the only mineral from which chymists are wont fruitlessly to attempt the separating of their three principles. I know indeed (continues Eleutherius) that the learned Sennertus, even in that book where he takes not upon him to play the advocate for the chymists, but the umpier betwixt them and the peripateticks, expresses himself roundly, thus; " Salem omnibus inesse (mixtis scilicet) et ex iis fieri posse omnibus in resolutionibus chymicis versatis notissimum est." And in the next page, " Quod de sale dixi," saies he, " idem de sulphure dici potest: " but by his favour I must see very good proofs, before I believe such general assertions, how boldly soever made; and he that would convince me of their truth, must first teach me some true and practicable way of separating salt and sulphur from gold, silver, and those many different sorts of stones, that a violent fire does not bring to lime, but to fusion; and not only I, for my own part, never saw any of those newly named bodies so resolved; but Helmont, who was much better versed in the chymical anatomizing of bodies than either Sennertus or I, has somewhere this resolute passage; " Scio (saies he) ex arena, silicibus et saxis, non calcariis, numquam sulphur aut mercurium trahi posse; ' nay Quercetanus himself, though the grand stickler for the *tria prima*, has this confession of the irresolubleness of diamonds; " Adamas (saith he) omnium factus lapidum solidissimus ac durissimus ex arctissima videlicet trium principiorum unione ac cohærentia, quæ nulla arte separationis in solutionem principiorum suorum spiritualium disjungi potest." And indeed, pursues Eleutherius, I was not only glad but somewhat surprized to find you

inclined to admit that there may be a sulphur and a
running mercury drawn from gold; for unless you do
(as your expression seemed to intimate) take the word
sulphur in a very loose sence, I must doubt whether our
chymists can separate a sulphur from gold: for when I
saw you make the experiment that I suppose invited you
to speak as you did, I did not judge the golden tincture
to be the true principle of sulphur extracted from the
body, but an aggregate of some such highly coloured
parts of the gold, as a chymist would have called a *sulphur
incombustible*, which in plain English seems to be little
better than to call it a sulphur and no sulphur. And as
for metalline mercuries, I had not wondred at it, though
you had expressed much more severity in speaking of
them: for I remember that having once met an old and
famous artist, who had long been (and still is) chymist
to a great monarch, the repute he had of a very honest
man invited me to desire him to tell me ingenuously
whether or no among his many labours, he had ever really
extracted a true and running mercury out of metalls; to
which question he freely replyed, that he had never
separated a true mercury from any metal; nor had ever
seen it really done by any man else. And though gold
is, of all metalls, that, whose mercury chymists have most
endeavoured to extract, and which they do the most brag
they have extracted; yet the experienced Angelus Sala,
in his spagyrical account of the seven terrestrial planets
(that is the seven metalls) affords us this memorable
testimony, to our present purpose; " Quanquam (saies he)
etc. experientia tamen (quam stultorum magistram
vocamus) certe comprobavit, mercurium auri adeo fixum,
maturum, et arcte cum reliquis ejusdem corporis
substantiis conjungi, ut nullo modo retrogredi possit."
To which he sub-joynes that he himself had seen much
labour spent upon that design, but could never see any
such mercury produced thereby. And I easily believe
what he annexes; " *that he had often seen detected many
tricks and impostures of cheating alchymists.* For, the
most part of those that are fond of such charlatans, being
unskilful or credulous, or both, 'tis very easie for such as

have some skill, much craft, more boldness, and no
conscience, to impose upon them; and therefore, though
many professed alchymists, and divers persons of quality
have told me that they have made or seen the mercury of
gold, or of this or that other metal; yet I have been still
apt to fear that either these persons have had a design
to deceive others; or have had not skill and circumspec-
tion enough to keep themselves from being deceived.

You recall to my mind (saies Carneades) a certain
experiment I once devised, innocently to deceive some
persons and let them and others see how little is to be built
upon the affirmation of those that are either unskilfull or
unwary, when they tell us they have seen alchymists make
the mercury of this or that metal; and to make this the
more evident, I made my experiment much more slight,
short and simple, than the chymists usuall processes to
extract metalline mercuries; which operations being
commonly more elaborate and intricate, and requiring
a much more longer time, give the alchymists a greater
opportunity to cozen, and consequently are more obnoxious to the spectators suspition. And that wherein
I endeavoured to make my experiment look the more like
a true analysis, was, that I not only pretended as well as
others to extract a mercury from the metal I wrought
upon, but likewise to separate a large proportion of
manifest and inflamable sulphur. I take then, of the
filings of copper, about a drachme or two; of common
sublimate, powdered, the like weight; and sal armoniack
near about as much as of sublimate; these three being well
mingled together I put into a small vial with a long neck,
or, which I find better, into a glass urinall, which (having
first stopped it with cotton) to avoid the noxious fumes,
I approach by degrees to a competent fire of well kindled
coals, or (which looks better, but more endangers the
glass) to the flame of a candle; and after a while the
bottom of the glass being held just upon the kindled coals,
or in the flame, you may in about a quarter of an hour,
or perchance in halfe that time, perceive in the bottom
of the glass some running mercury; and if then you take
away the glass and break it, you shall find a parcel of

quicksilver, perhaps altogether, and perhaps part of it in the pores of the solid mass; you shall find too, that the remaining lump being held to the flame of the candle will readily burn with a greenish flame, and after a little while (perchance presently) will in the air acquire a greenish blew, which being the colour that is ascribed to copper, when its body is unlocked, 'tis easie to perswade men that this is the true sulphur of Venus, especially since not only the salts may be supposed partly to be flown away, and partly to be sublimed to the upper part of the glass, whose inside (will commonly appear whitened by them) but the metal seems to be quite destroyed, the copper no longer appearing in a metalline forme, but almost in that of a resinous lump; whereas indeed the case is only this, that the saline parts of the sublimate together with the sal armoniack, being excited and actuated by the vehement heat, fall upon the copper, (which is a metal they can more easily corrode, than silver) whereby the small parts of the mercury being freed from the salts that kept them asunder, and being by the heat tumbled up and down after many occursions, they convene into a conspicuous mass of liquor; and as for the salts, some of the more volatile of them subliming to the upper part of the glass, the others corrode the copper, and uniting themselves with it do strangely alter and disguise its metallick form, and compose with it a new kind of concrete inflamable like sulphur; concerning which I shall not now say anything, since I can referr you to the diligent observations which I remember Mr. Boyle has made concerning this odde kind of verdigrease. But continues Carneades smiling, you know I was not cut out for a mountebank, and therefore I will hasten to resume the person of a sceptick, and take up my discourse where you diverted me from prosecuting it.

In the next place, then, I consider, that, as there are some bodies which yield not so many as the three principles; so there are many others, that in their resolution exhibit more principles than three; and that therefore the ternary number is not that of the universal and adequate principles of bodies. If you allow of the discourse I lately made you, touching the primary associations of the small particles of matter, you will scarce think it improbable, that of such elementary corpuscles there may be more sorts than either three, or four, or five. And if you will grant, what will scarce be denied, that corpuscles of a compounded nature may in all the wonted examples of chymists pass for elementary, I see not why you should think it impossible, that as *aqua fortis*, or *aqua regis* will make a separation of colliquated silver and gold, though the fire cannot; so there may be some agent found out so subtile and so powerfull, at least in respect of those particular compounded corpuscles, as to be able to resolve them into those more simple ones, whereof they consist, and consequently encrease the number of the distinct substances, whereinto the mixt body has been hitherto thought resoluble. And if that be true, which I recited to you a while ago out of Helmont concerning the operations of the alkahest, which divides bodies into other distinct substances, both as to number and nature, than the fire does; it will not a little countenance my conjecture. But confining ourselves to such waies of analyzing mixed bodies, as are already not unknown to chymists, it may without absurdity be questioned, whether besides those grosser elements of bodies, which they call salt sulphur and mercury, there may not be ingredients of a more subtile nature, which being extreamly little, and not being in themselves visible, may escape unheeded at the junctures of the destillatory vessels, though never so carefully luted. For let me observe to you one thing, which though not taken notice of by chymists, may be a notion of good use in divers cases to a naturalist, that we may well suspect, that there may be severall sorts of bodies, which are not immediate objects of any one of our senses; since we see, that not only those little corpuscles that issue out of the loadstone, and perform the wonders for which it is justly admired; but the effluviums of amber, jet, and other electricall concretes, though by their effects upon the particular bodies disposed to receive their action, they seem to fall under the cognizance of our sight, yet do they not as electrical immediately affect any of our senses, as do the bodies, whether minute or greater, that we see, feel, taste, etc. But, (continues Carneades) because you may expect I should, as the chymists do, consider only the sensible ingredients of mixt bodies, let us now see, what experience will, even as to these, suggest to us.

It seems then questionable enough, whether from grapes variously ordered there may not be drawn more distinct substances by the help of the fire, than from most other mixt bodies. For the grapes themselves being dryed into raisins and distilled, will (besides alcali, phlegm, and earth) yeeld a considerable quantity of an empyreumatical oyle, and a spirit of a very different nature from that of wine. Also the unfermented juice of grapes affords other distilled liquors than wine doth. The juice of grapes after fermentation will yeeld a *spiritus ardens ;* which if competently rectifyed will all burn away without leaving anything remaining. The same fermented juice degenerating into vinegar, yeelds an acid and corroding spirit. The same juice tunned up, armes itself with tartar; out of which may be separated, as out of other bodies, phlegme, spirit, oyle, salt and earth: not to mention what substances may be drawn from the vine itselfe, probably differing from those which are separated from tartar, which is a body by itself, that has few resemblers in the world. And I will further consider that what force soever you will allow this instance, to evince that there are some bodies that yeeld more elements than others, it can scarce be denyed but that the major part of bodies that are divisible into elements yeeld more than three. For, besides those which the chymists are pleased to name hypostatical, most bodies contain two others, phlegme and earth, which concurring as well as the rest to the constitution of mixts, and being as generally, if not more, found in their analysis, I see no sufficient cause why they should be excluded from the number of elements. Nor will it suffice to object, as the Paracelsians are wont to do, that the *tria prima* are the most useful elements, and the earth and water but worthless and unactive; for elements being called so in relation to the constituting of mixt bodies, it should be upon the account of its ingrediency, not of its use, that anything should be affirmed or denied to be an element: and as for the pretended uselessness of earth and water, it would be considered that usefulness, or the want of it, denotes only a respect or relation to us; and therefore the presence, or absence of it, alters not the intrinsick nature of the thing. The hurtful teeth of vipers are for ought I know useless to us, and yet are not to be denyed to be parts of their bodies; and it were hard to shew of what greater use to us, than phlegme and earth, are those undiscerned stars, which our new telescopes discover to us, in many blanched places of the sky; and yet we cannot but acknowledge them constituent and considerably great parts of the universe. Besides that whether or no the phlegm and earth be immediately useful, but necessary to constitute the body whence they are separated; and consequently, if the mixt body be not useless to us, those constituent parts, without which it could not have been that mixt body, may be said not to be unuseful to us: and though the earth and water be not so conspicuously operative (after separation) as the other three more active principles, yet in this case it will not be amiss to remember the lucky fable of Menenius Agrippa, of the dangerous sedition of the hands and legs, and other more busie parts of the body, against the seemingly unactive stomack. And to this case also we may not unfitly apply that reasoning of an apostle, to another purpose; " If the ear shall say, because I am not the eye, I am not of the body; is it therefore not of the body? If the whole body were eye, where were the hearing? If the whole were for hearing, where the smelling? In a word, since earth and water appear, as clearly and as generally as the other principles upon the resolution of bodies, to be the ingredients whereof they are made up; and since they are useful (if not immediately to us, or rather to physitians) to the bodies they constitute, and so though in somewhat a remoter way, are serviceable to us; to exclude them out of the number of elements, is not to imitate nature.

And on this occasion I cannot but take notice, that

whereas the great argument which the chymists are wont
to employ to vilify earth and water, and make them be
looked upon as useless and unworthy to be reckoned
among the principles of mixt bodies, is, that they are not
endowed with specifick properties, but only with elementary qualities; of which they use to speak very slightingly,
as of qualities contemptible and unactive: I see no
sufficient reason for this practice of the chymists: for
'tis confessed that heat is an elementary quality, and yet
that an almost innumerable company of considerable
things are performed by heat, is manifest to them that
duly consider the various phænomena wherein it intervenes as a principall actor; and none ought less to ignore
or distrust this truth than a chymist. Since almost all
the operations and productions of his art are performed
chiefly by the means of heat. And as for cold itself, upon
whose account they so despise the earth and water, if
they please to read in the voyages of our English and
Dutch navigators in Nova Zembla and other northern
regions what stupendous things may be effected by cold,
they would not perhaps think it so despicable. And not
to repeat what I lately recited to you out of Paracelsus
himself, who by the help of an intense cold teaches to
separate the quintessence of wine; I will only now
observe to you, that the conservation of the texture of
many bodies both animate and inanimate, does so much
depend upon the convenient motion both of their own
fluid and looser parts, and of the ambient bodies, whether
air, water, etc. that not only in humane bodies we see
that the immoderate or unseasonable coldness of the air
(especially when it finds such bodies overheated) does
very frequently discompose the oeconomie of them, and
occasion variety of diseases; but in the solid and durable
body of iron itself, in which one would not expect that
suddain cold should produce any notable change, it may
have so great an operation, that if you take a wire, or
other slender piece of steel, and having brought it in the
fire to a white heat, you suffer it afterwards to cool
leasurely in the air, it will when it is cold be much of the
same hardness it was of before. Whereas if as soon as

you remove it from the fire, you plunge it into cold water, it will upon the suddain refrigeration acquire a very much greater hardness than it had before; nay, and will become manifestly brittle. And that you may not impute this to any peculiar quality in the water, or other liquor, or unctuous matter, wherein such heated steel is wont to be quenched that it may be tempered; I know a very skilful tradesman, that divers times hardens steel by suddenly cooling it in a body that is neither a liquor, nor so much as moist. A tryal of that nature I remember I have seen made. And however by the operation that water has upon steel quenched in it, whether upon the account of its coldness and moisture, or upon that of any other of its qualities, it appears, that water is not alwaies so inefficacious and contemptible a body, as our chymists would have it pass for. And what I have said of the efficacy of cold and heat, might perhaps be easily enough carried further by other considerations and experiments; were it not that having been mentioned only upon the by, I must not insist on it, but proceed to another subject.

But, (pursues Carneades) though I think it evident, that earth and phlegme are to be reckoned among the elements of most animal and vegetable bodies, yet 'tis not upon that account alone, that I think divers bodies resoluble into more substances than three. For there are two experiments, that I have sometimes made to shew, that at least some mixts are divisible into more distinct substances than five. The one of these experiments, though 'twill be more seasonable for me to mention it fully anon, yet in the meantime, I shall tell you thus much of it, that out of two distilled liquors which pass for elements of the bodies whence they are drawn, I can without addition make a true yellow and inflamable sulphur, notwithstanding that the two liquors remain afterwards distinct. Of the other experiment, which perhaps will not be altogether unworthy your notice, I must now give you this particular account. I had long observed, that by the destillation of divers woods, both in ordinary, and some unusuall sorts of vessels, the copious spirit that came over, had besides a strong taste, to be met with in the empyreumatical spirits of many other bodies, an acidity almost like that of vinegar: wherefore I suspected, that though the sowrish liquor distilled, for instance, from box-wood, be lookt upon by chymists as barely the spirit of it, and therefore as one single element or principle; yet it does really consist of two differing substances, and may be divisible into them; and consequently, that such woods and other mixts as abound with such a vinegar, may be said to consist of one element or principle, more than the chymists as yet are aware of, wherefore bethinking myself, how the separation of these two spirits might be made, I quickly found, that there were several waies of compassing it. But that of them which I shall at present mention was this, Having distilled a quantity of box-wood *per se*, and slowly rectifyed the sowrish spirit, the better to free it both from oyle and phlegme, I cast into this rectifyed liquor a convenient quantity of powdered coral, expecting that the acid part of the liquor, would corrode the coral, and being associated with it would be so retained by it, that the other part of the liquor, which was not of an acid nature, nor fit to fasten upon the corals, would be permitted to ascend alone. Nor was I deceived in my expectation; for having gently abstracted the liquor from the corals, there came over a spirit of a strong smell, and of a taste very piercing but without any sowrness; and which was in diverse qualities manifestly different, not only from a spirit of vinegar, but from some spirit of the same wood, that I purposely kept by me without depriving it of its acid ingredient. And to satisfy you, that these two substances were of a very differing nature, I might informe you of several tryals that I made, but must not name some of them, because I cannot do so without making some unseasonable discoveries. Yet this I shall tell you at present that the sowre spirit of *box*, not only would, as I just now related, dissolve corals, which the other would not fasten on, but being poured upon salt of tartar would immediately boyle and hiss, whereas the other would lye quietly upon it. The acid spirit poured upon minium made a sugar of lead, which I did not find the other to do; some drops of this penetrant spirit being mingled with some drops of the blew syrup of violets seemed rather to dilute than otherwise alter the colour; whereas the acid spirit turned the syrup of a reddish colour, and would probably have made it of as pure a red, as acid salts are wont to do, had not its operation been hindered by the mixture of the other spirit. A few drops of the compound spirit being shaken into a pretty quantity of the infusion of *lignum nephriticum,* presently destroyed all the blewish colour, whereas the other spirit would not take it away. To all which it might be added, that having for tryals sake poured fair water upon the corals that remained in the bottom of the glass wherein I had rectifyed the double spirit (if I may so call it) that was first drawn from the box, I found according to my expectation that the acid spirit had really dissolved the corals and had coagulated with them. For by the affusion of fair water, I obtained a solution, which (to note that singularity upon the by) was red, whence the water being evaporated, there remained a soluble substance much like the ordinary salt of coral, as chymists are pleased to call that magistery of corals, which they make by dissolving them in common spirit of vinegar, and abstracting the *menstruum ad siccitatem.* I know not whether I should subjoyne, on this occasion, that the simple spirit of box, if chymists will have it therefore saline because it has a strong taste, will furnish us with a new kind of saline bodies, differing from those hitherto taken notice of. For whereas of the three chief sorts of salts, the acid, the alcalizate, and the sulphureous, there is none that seems to be friends with both the other two, as I may, ere it be long, have occasion to shew; I did not find but that the simple spirit of box did agree very well (at least as farr as I had occasion to try it) both with the acid and the other salts. For though it would lye very quiet with salt of tartar, spirit of urine, or other bodies, whose salts were either of an alcalizate or fugitive nature; yet did not the mingling of oyle of vitriol itself produce any hissing or effervescence, which you know is wont to ensue upon the affusion of that highly acid liquor upon either of the bodies newly mentioned.

I think myself, (saies Eleutherius) beholden to you, for this experiment; not only because I foresee you will make it helpful to you in the enquiry you are now upon, but because it teaches us a method, whereby we may prepare a numerous sort of new spirits, which though more simple than any that are thought elementary, are manifestly endowed with peculiar and powerful qualities, some of which may probably be of considerable use in physick, as well alone as associated with other things; as one may hopefully guess by the redness of that solution your sowre spirit made of corals, and by some other circumstances of your narrative. And suppose (pursues Eleutherius) that you are not so confined, for the separation of the acid parts of these compound spirits from the other, to employ corals; but that you may as well make use of any alcalizate salt, or of pearls, or crabs eyes, or any other body, upon which common spirit of vinegar will easily work, and, to speak in an Helmontian phrase, exantlate itself.

I have not yet tryed, (saies Carneades) of what use the mentioned liquors may be in physick, either as medicines or as menstruums: but I could mention now (and may another time) divers of the tryals that I made to satisfy myself of the difference of these two liquors. But that, as I allow your thinking what you newly told me about corals, I presume you will allow me, from what I have said already, to deduce this corollary; that there are divers compound bodies, which may be resolved into four such differing substances, as may as well merit the name of principles, as those to which the chymists freely give it. For since they scruple not to reckon that which I call the compound spirit of box, for the spirit, or as others would have it, the mercury of that wood, I see not, why the acid liquor, and the other, should not each of them, especially that last named, be lookt upon as more worthy to be called an elementary principle; since it must needs be of a more simple nature than the liquor, which was found to be divisible into that, and the acid spirit. And this further use (continues Carneades) may be made of our experiment to my present purpose, that it may give us a rise to suspect, that since a liquor reputed by the chymists to be, without dispute, homogeneous, is by so slight a way divisible into two distinct and more simple ingredients, some more skilful or happier experimenter than I may find a way either further to divide one of these spirits, or to resolve some or other, if not all, of those other ingredients of mixt bodies, that have hitherto passed among chymists for their elements or principles.

THE FOURTH PART

AND thus much (saies Carneades) may suffice to be said of the *number* of the distinct substances separable from mixt bodies by the fire: wherefore I now proceed to consider the *nature* of them, and shew you, that though they seem *homogeneous* bodies, yet have they not the purity and simplicity that is requisite to elements. And I should immediately proceed to the proof of my assertion, but that the confidence wherewith chymists are wont to call each of the substances we speak of by the name of sulphur or mercury, or the other of the hypostatical principles, and the intolerable ambiguity they allow themselves in their writings and expressions, makes it necessary for me in order to the keeping you either from mistaking me, or thinking I mistake the controversie, to take notice to you and complain of the unreasonable liberty they give themselves of playing with names at pleasure. And indeed if I were obliged in this dispute, to have such regard to the phraseology of each particular chymist, as not to write anything which this or that author may not pretend, not to contradict this or that sence, which he may give us as occasion serves to his ambiguous expressions, I should scarce know how to dispute, nor which way to turn myself. For I find that even eminent writers (such as Raymund Lully, Paracelsus and others) do so abuse the termes they employ, that as they will now and then give divers things, one name; so they will oftentimes give one thing, many names; and some of them (perhaps) such, as do much more properly signifie some distinct body of another kind; nay even in technical words or termes of art, they refrain not from this confounding liberty; but will, as I have observed, call the same substance, sometimes the sulphur, and sometimes the mercury of a body. And now I speak of mercury, I cannot but take notice, that the descriptions they give us of that principle or ingredient of mixt bodies, are so intricate, that even those that have endeavoured to polish and illustrate the notions of the chymists, are fain to confess that they know not what to make of it either by ingenuous acknowledgments, or descriptions that are not intelligible.

I must confess (saies Eleutherius) I have, in the reading of Paracelsus and other chymical authors, been troubled to find, that such hard words and equivocal expressions, as you justly complain of, do even when they treat of principles, seem to be studiously affected by those writers; whether to make themselves to be admired by their readers, and their art appear more venerable and mysterious, or (as they would have us think) to conceal from them a knowledge themselves judge inestimable.

But whatever (saies Carneades) these men may promise themselves from a canting way of delivering the principles of nature, they will find the major part of knowing men so vain, as when they understand not what they read, to conclude, that it is rather the writers fault than their own. And those that are so ambitious to be admired by the vulgar, that rather than go without the admiration of the ignorant they will expose themselves to the contempt of the learned, those shall, by my consent, freely enjoy their option. As for the mystical writers scrupling to communicate their knowledge, they might less to their own disparagement, and to the trouble of their readers, have concealed it by writing no books, than by writing bad ones. If Themistius were here, he would not stick to say, that chymists write thus darkly, not because they think their notions too precious to be explained, but because they fear that if they were explained, men would discern, that they are farr from being precious. And indeed, I fear that the chief reason why chymists have written so obscurely of their three principles, may be, that not having clear and distinct notions of them themselves, they cannot write otherwise than confusedly of what they but confusedly apprehend: not to say that divers of them, being conscious to the invalidity of their doctrine, might well enough discerne that they could scarce keep themselves from being confuted, but by keeping themselves from being clearly understood. But though much may be said to excuse the chymists when they write darkly, and ænigmatically, about the preparation of their elixir, and some few other grand arcana, the divulging of which they may upon grounds plausible enough esteem unfit; yet when they pretend to teach the general principles of natural philosophers, this equivocal way of writing is not to be endured. For in such speculative enquiries, where the naked knowledge of the truth is the thing principally aimed at, what does he teach me worth thanks that does not, if he can, make his notion intelligible to me, but by mystical termes, and ambiguous phrases darkens what he should clear up; and makes me add the trouble of guessing at the sence of what he equivocally expresses, to that of examining the truth of what he seems to deliver. And if the matter of the philosophers stone, and the manner of preparing it, be such mysteries as they would have the world believe them, they may write intelligibly and clearly of the principles of mixt bodies in general, without discovering what they call the great work. But for my part (continues Carneades) what my indignation at this unphilosophical way of teaching principles has now extorted from me, is meant chiefly to excuse myself, if I shall hereafter oppose any particular opinion or assertion, that some follower of Paracelsus or any eminent artist may pretend not to be his masters. For, as I told you long since, I am not obliged to examine private men's writings, (which were a labour as endless as unprofitable) being only engaged to examine those opinions about the *tria prima*, which I find those chymists I have met with to agree in most: and I doubt not but my arguments against their doctrine will be in great part easily enough applicable even to those private opinions, which they do not so directly and expressly oppose. And indeed, that which I am now entering upon being the consideration of the things themselves whereinto spagyrists resolve mixt bodies by the fire, if I can shew that these are not of an elementary nature, it will be no great matter what names these or those chymists have been pleased to give them. And I question not that to a wise man, and consequently to Eleutherius, it will be lesse considerable to know, what men have thought of things, than what they should have thought.

In the fourth and last place, then, I consider, that as generally as chymists are wont to appeal to experience, and as confidently as they use to instance the several substances separated by the fire from a mixt body, as a sufficient proof of their being its component elements: yet those differing substances are many of them farr enough from elementary simplicity, and may be yet looked upon as mixt bodies, most of them also retaining, somewhat at least, if not very much, of the nature of those concretes whence they were forced.

I am glad (saies Eleutherius) to see the vanity or envy of the canting chymists thus discovered and chastised; and I could wish, that learned men would conspire together to make these deluding writers sensible, that they must no longer hope with impunity to abuse the world. For whilst such men are quietly permitted to publish books with promising titles, and therein to assert what they please, and contradict others, and even themselves as they please, with as little danger of being confuted as of being understood, they are encouraged to get themselves a name, at the cost of the readers, by finding that intelligent men are wont for the reason newly mentioned, to let their books and them alone: and the ignorant and credulous (of which the number is still much greater than that of the other) are forward to admire most what they least understand. But if judicious men skilled in chymical affaires shall once agree to write clearly and plainly of them, and thereby keep men from being stunned, as it were, or imposed upon by dark or empty words; 'tis to be hoped that these men finding that they can no longer write impertinently and absurdly, without being laughed at for doing so, will be reduced either to write nothing, or books that may teach us something, and not rob men, as formerly, of invaluable time; and so ceasing to trouble the world with riddles or impertinencies, we shall either by their books receive an advantage, or by their silence escape an inconvenience.

But after all this is said (continues Eleutherius) it may be represented in favour of the chymists, that, in one regard the liberty they take in using names, if it be excusable at any time, may be more so when they speak of the substances whereinto their analysis resolves mixt bodies: since as parents have the right to name their own children, it has ever been allowed to the authors of new inventions, to impose names upon them. And therefore the subjects we speak of being so the productions of the chymists art, as not to be otherwise, but by it, obtainable; it seems but equitable to give the artists leave to name them as they please: considering also that none are so fit and likely to teach us what those bodies are, as they to whom we owed them.

I told you already (saies Carneades) that there is great difference betwixt the being able to make experiments, and the being able to give a philosophical account of them. And I will not now add, that many a mine-digger may meet, whilst he follows his work, with a gemm or a mineral which he knowes not what to make of, till he shewes it a jeweller or a mineralist to be informed what it is. But that which I would rather have here observed is, that the chymists I am now in debate with have given up the liberty you challenged for them, of using names at pleasure, and confined themselves by their descriptions, though but such as they are, of their principles; so that although they might freely have called anything their analysis presents them with, either sulphur, or mercury, or gas, or blas, or what they pleased; yet when they have told me that sulphur (for instance) is a primogeneal and simple body, inflamable, odorous, etc. they must give me leave to disbelieve them, if they tell me that a body that is either compounded or uninflamable is such a sulphur; and to think they play with words, when they teach that gold and some other minerals abound with an incombustible sulphur, which is as proper an expression, as a sun-shine night, or fluid ice.

But before I descend to the mention of particulars

belonging to my fourth consideration, I think it convenient
to premise a few generals; some of which I shall the less
need to insist on at present, because I have touched on
them already.

And first I must invite you to take notice of a certain
passage in Helmont;[1] which though I have not found
much heeded by his readers, he himself *mentions* as a
notable thing, and I take to be a very considerable one; for
whereas the distilled oyle of *oyle-olive*, though drawn *per se*
is (as I have tryed) of a very sharp and fretting quality,
and of an odious taste, he tells us that simple oyle being
only digested with Paracelsus's *sal circulatum*, is reduced
into dissimilar parts, and yeelds a sweet oyle, very differing
from the oyle distilled, from sallet oyle; as also that by
the same way there may be separated from wine a very
sweet and gentle spirit, partaking of a far other and
nobler quality than that which is immediately drawn by
distillation and called *dephlegmed aqua vitæ*, from whose
acrimony this other spirit is exceedingly remote, although
the *sal circulatum* that makes these anatomies be separated
from the analyzed bodies, in the same weight and with
the same qualities it had before; which affirmation of
Helmont if we admit to be true, we must acknowledge
that there may be a very great disparity betwixt bodies
of the same denomination (as several oyles, or several
spirits) separable from compound bodies: for, besides the
differences I shall anon take notice of, betwixt those
distilled oyles that are commonly known to chymists, it
appears by this, that by means of the *sal circulatum*, there
may be quite another sort of oyles obtained from the same
body; and who knowes but that there may be yet other
agents found in nature, by whose help there may, whether
by transmutation or otherwise, be obtained from the
bodies vulgarly called mixt, oyles or other substances,
differing from those of the same denomination, known
either to vulgar chymists, or even to Helmont himself:
but for fear you should tell me, that this is but a conjecture grounded upon another man's relation, whose
truth we have not the means to experiment, I will not

[1] Helmont, *Aura vitalis*, p. 725.

insist upon it; but leaving you to consider of it at leasure, I shall proceed to what is next.

Secondly, then, if that be true which was the opinion of Leucippus, Democritus, and other prime anatomists of old, and is in our dayes revived by no mean philosophers; namely, that our culinary fire, such as chymists use, consists of swarmes of little bodies swiftly moving, which by their smallness and motion are able to permeate the sollidest and compactest bodies, and even glass itself; if this (I say) be true, since we see that in flints and other concretes, the fiery part is incorporated with the grosser, it will not be irrational to conjecture, that multitudes of these fiery corpuscles, getting in at the pores of the glass, may associate themselves with the parts of the mixt body whereon they work, and with them constitute new kinds of compound bodies, according as the shape, size, and other affections of the parts of the dissipated body happen to dispose them, in reference to such combinations; of which also there may be the greater number; if it be likewise granted that the corpuscles of the fire, though all exceeding minute, and very swiftly moved, are not all of the same bigness, nor figure: and if I had not weightier considerations to discourse to you of, I could name to you, to countenance what I have newly said, some particular experiments by which I have been deduced to think, that the particles of an open fire working upon some bodies may really associate themselves therewith, and add to the quantity. But because I am not sure, that when the fire works upon bodies included in glasses, it does it by a reall trajection of the fiery corpuscles themselves, through the substance of the glass, I will proceed to what is next to be mentioned.

I could (saies Eleutherius) help you to some proofs, whereby I think it may be made very probable, that when the fire acts immediately upon a body, some of its corpuscles may stick to those of the burnt body, as they seem to do in quicklime, but in greater numbers and more permanently. But for fear of retarding your progress, I shall desire you to deferr this enquiry till another time, and proceed as you intended.

You may then in the next place (saies Carneades) observe with me, that not only there are some bodies, as gold, and silver, which do not by the usual examens, made by fire, discover themselves to be mixt; but if (as you may remember I formerly told you) it be a decompound body that is dissipable into several substances, by being exposed to the fire it may be resolved into such as are neither elementary, nor such as it was upon its last mixture compounded of; but into new kinds of mixts. Of this I have already given you some examples in sope, sugar of lead, and vitriol. Now if we shall consider that there are some bodies, as well natural, (as that I last named) as factitious, manifestly decompounded; that in the bowells of the earth nature may, as we see she sometimes does, make strange mixtures; that animals are nourished with other animals and plants; and, that these themselves have almost all of them their nutriment and growth, either from a certain nitrous juice harboured in the pores of the earth, or from the excrements of animalls, or from the putrifyed bodies, either of living creatures or vegetables, or from other substances of a compounded nature; if, I say, we consider this, it may seem probable, that there may be among the works of nature (not to mention those of art) a greater number of decompound bodies, than men take notice of; and indeed, as I have formerly also observed, it does not at all appear, that all mixtures must be of elementary bodies; but it seems farr more probable, that there are divers sorts of compound bodies, even in regard of all or some of their ingredients, considered antecedently to their mixture. For though some seem to be made up by the immediate coalitions of the elements, or principles themselves, and therefore may be called *prima mista*, or *mista primaria;* yet it seems that many other bodies are mingled (if I may so speak) at the second hand, their immediate ingredients being not elementary, but these primary mixt newly spoken of; and from divers of those secondary sorts of mixts may result, by a further composition, a third sort, and so onwards. Nor is it improbable, that some bodies are made up of mixt bodies, not all of the same order, but of several; as (for instance) a concrete may consist of ingredients, whereof the one may have been a primary, the other a secondary mixt body; (as I have in native cinnaber, by my way of resolving it, found both that courser part that seems more properly to be oar, and a combustible sulphur, and a running mercury): or perhaps without any ingredient of this latter sort, it may be composed of mixt bodies, some of them of the first, and some of the third kind; and this may perhaps be somewhat illustrated by reflecting upon what happens in some chymical preparations of those medicines which they call their *Bezoardicum's*. For first, they take antimony and iron, which may be looked upon as *prima mista ;* of these they compound a starry *regulus*, and to this they add according to their intention, either gold, or silver, which makes with it a new and further composition. To this they add sublimate, which is itself a decompound body, (consisting of common quicksilver, and divers salts united by sublimation into a chrystalline substance) and from this sublimate, and the other metalline mixtures, they draw a liquor, which may be allowed to be of a yet more compounded nature. If it be true, as chymists affirm it, that by this art some of the gold or silver mingled with the *regulus* may be carried over the helme with it by the sublimate; as indeed a skilfull and candid person complained to me a while since, that an experienced friend of his and mine, having by such a way brought over a great deal of gold, in hope to do something further with it, which might be gainful to him, has not only missed of his aim, but is unable to recover his volatilized gold out of the antimonial butter, wherewith it is strictly united.

Now (continues Carneades) if a compound body consist of ingredients that are not merely elementary; it is not hard to conceive, that the substances into which the fire dissolves it, though seemingly homogeneous enough, may be of a compounded nature, those parts of each body that are most of kin associating themselves into a compound of a new kind. As when (for example sake) I have caused vitriol and sal armoniack, and salt petre to be mingled and distilled together, the liquor that came over manifested itself not to be either spirit of nitre, or of sal armoniack, or of vitrioll.   For none of these would dissolve crude gold, which yet my liquor was able readily to do; and thereby manifested itself to be a new compound, consisting at least of spirit of nitre, and sal armoniack, (for the latter dissolved in the former, will work on gold) which nevertheless are not by any known way separable, and consequently would not pass for a mixt body, if we ourselves did not, to obtain it, put and distill together divers concretes, whose distinct operations were known beforehand.   And, to add on this occasion the experiment I lately promised you, because it is applicable to our present purpose, I shall acquaint you, that suspecting the common oyle of vitrioll not to be altogether such a simple liquor as chymists presume it, I mingled it with an equal or a double quantity (for I tryed the experiment more than once) of common oyle of turpentine, such as together with the other liquor I bought at the drugsters. And having carefully (for the experiment is nice, and somewhat dangerous) distilled the mixture in a small glass retort, I obtained according to my desire (besides, the two liquors I had put in) a pretty quantity of a certaine substance, which sticking all about the neck of the retort discovered itself to be sulphur, not only by a very strong sulphureous smell, and by the colour of brimstone; but also by this, that being put upon a coal, it was immediately kindled, and burned like common sulphur.   And of this substance I have yet by me some little parcells, which you may command and examine when you please.   So that from this experiment I may deduce either one, or both of these propositions, that a real sulphur may be made by the conjunction of two such substances as chymists take for elementary, and which did not either of them apart appear to have any such body in it; or that oyle of vitrioll though a distilled liquor, and taken for part of the saline principle of the concrete that yeelds it, may yet be so compounded a body as to contain, besides its saline part, a sulphur like common brimstone, which would hardly be itself a simple or uncompounded body.

I might (pursues Carneades) remind you, that I formerly

represented it, as possible, that as there may be more elements than five, or six; so the elements of one body may be different from those of another; whence it would follow, that from the resolution of decompound bodies, there may result mixts of an altogether new kind, by the coalition of elements that never perhaps convened before. I might, I say, mind you of this, and add divers things to this second consideration; but for fear of wanting time I willingly pretermit them to pass on to the third, which is this, that the fire does not alwaies barely resolve or take asunder, but may also after a new manner mingle and compound together the parts (whether elementary or not) of the body dissipated by it.

This is so evident, (saies Carneades) in some obvious examples, that I cannot but wonder at their supineness that have not taken notice of it. For when wood being burnt in a chimney is dissipated by the fire into smoake and ashes, that smoake composes soot, which is so far from being any one of the principles of the wood, that (as I noted above) you may by a further analysis separate five or six distinct substances from it. And as for the remaining ashes, the chymists themselves teach us, that by a further degree of fire they may be indissolubly united into glass. 'Tis true, that the analysis which the chymists principally build upon is made, not in the open air, but in close vessels; but however, the examples lately produced may invite you shrewdly to suspect, that heat may as well compound as dissipate the parts of mixt bodies: and not to tell you, that I have known a vitrification made even in close vessels, I must remind you that the flowers of antimony, and those of sulphur, are very mixed bodies, though they ascend in close vessels: and that 'twas in stopt glasses that I brought up the whole body of camphire. And whereas it may be objected that all these examples are of bodies forced up in a dry, not a fluid forme, as are the liquors wont to be obtained by distillation; I answer, that besides 'tis possible, that a body may be changed from consistent to fluid, or from fluid to consistent, without being otherwise much altered, as may appear by the easiness wherewith in winter, without any addition or separation of visible ingredients, the same substance may be quickly hardened into brittle ice, and thawed again into fluid water; besides this, I say it would be considered, that common quicksilver itself, which the eminentest chymists confess to be a mixt body, may be driven over the helme in its pristine forme of quicksilver, and consequently, in that of a liquor. And certainly 'tis possible that very compounded bodies may concurr to constitute liquors; since, not to mention that I have found it possible, by the help of a certain menstruum, to distill gold itself through a retort, even with a moderate fire: let us but consider what happens in butter of antimony. For if that be carefully rectifyed, it may be reduced into a very clear liquor; and yet if you cast a quantity of fair water upon it, there will quickly precipitate a ponderous and vomitive calx, which made before a considerable part of the liquor, and yet is indeed (though some eminent chymists would have it mercurial) an antimonial body carryed over and kept dissolved by the salts of the sublimate, and consequently a compounded one; as you may find, if you will have the curiosity to examine this white powder by a skilful reduction. And that you may not think that bodies as compounded, as flowers of brimstone, cannot be brought to concurr to constitute distilled liquors; and also that you may not imagine with divers learned men that pretend no small skill in chymistry, that at least no mixt body can be brought over the helme, but by corrosive salts, I am ready to shew you, when you please, among other waies of bringing over flowers of brimstone (perhaps I might add even mineral sulphurs) some, wherein I employ none but oleaginous bodies to make volatile liquors, in which not only the colour, but (which is a much surer mark) the smell and some operations manifest that there is brought over a sulphur that makes part of the liquor.

One thing more there is Eleutherius, (saies Carneades) which is so pertinent to my present purpose, that though I have touched upon it before, I cannot but on this occasion take notice of it. And it is this, that the qualities or accidents, upon whose account chymists are wont to

call a portion of matter by the name of mercury or some
other of their principles, are not such but that 'tis possible
as great (and therefore why not the like) may be produced
by such changes of texture, and other alterations, as the
fire may make in the small parts of a body. I have
already proved, when I discoursed of the second general
consideration, by what happens to plants nourished only
with fair water, and eggs hatched into chickens, that by
changing the disposition of the component parts of a body,
nature is able to effect as great changes in a parcell of
matter reputed similar, as those requisite to denominate
one of the *tria prima*. And though Helmont do somewhere wittily call the fire the destructor and the artificial
death of things; and although another eminent chymist
and physitian be pleased to build upon this, that fire
can never generate anything but fire; yet you will, I
doubt not, be of another mind, if you consider how many
new sorts of mixt bodies chymists themselves have produced by means of the fire: and particularly, if you
consider how that noble and permanent body, glass, is
not only manifestly produced by the violent action of the
fire, but has never, for ought we know, been produced any
other way. And indeed it seems but an inconsiderate
assertion of some Helmontians, that every sort of body
of a peculiar denomination must be produced by some
seminal power; as I think I could evince, if I thought it
so necessary, as it is for me to hasten to what I have
further to discourse. Nor need it much move us, that
there are some who look upon whatsoever the fire is
employed to produce, not as upon natural but artificial
bodies. For there is not alwaies such a difference as
many imagine betwixt the one and the other: nor is it
so easy as they think, clearly to assigne that which
properly, constantly, and sufficiently, discriminates them.
But not to engage myself in so nice a disquisition, it may
now suffice to observe, that a thing is commonly termed
artificial, when a parcel of matter is by the artificers hand,
or tools, or both, brought to such a shape or form, as he
designed beforehand in his mind: whereas in many of
the chymical productions the effect would be produced

whether the artificer intended it or no; and is oftentimes very much other than he intended or looket for; and the instruments employed, are not tools artificially fashioned and shaped, like those of tradesmen, for this or that particular work; but, for the most part, agents of nature's own providing, and whose chief powers of operation they receive from their own nature or texture, not the artificer. And indeed, the fire is as well a natural agent as seed: and the chymist that imployes it, does but apply natural agents and patients, who being thus brought together, and acting according to their respective natures, performe the work themselves; as apples, plums, or other fruit, are natural productions, though the garden bring and fasten together the sciens and the stock, and both water, and do perhaps divers other waies contribute to its bearing fruit.   But, to proceed to what I was going to say; you may observe with me, Eleutherius, that, as I told you once before, qualities sleight enough may serve to denominate a chymical principle.   For, when they anatomize a compound body by the fire, if they get a substance inflamable, and that will not mingle with water, that they presently call sulphur; what is sapid and dissoluble in water, that must passe for salt; whatsoever is fixed and indissoluble in water, that they name earth.   And I was going to add, that whatsoever volatile substance they know not what to make of, not to say, whatsoever they please, that they call mercury.   But that these qualities may either be produced, otherwise than by such as they call seminal agents, or may belong to bodies of a compounded nature, may be shewn, among other instances, in glass made of ashes, where the exceeding strong-tasted *alcalizate* salt joyning with the earth becomes insipid, and with it constitutes a body; which though also dry, fixt and indissoluble in water, is yet manifestly a mixt body; and made so by the fire itself.

And I remember to our present purpose, that Helmont, amongst other medicines that he commends, has a short process, wherein, though the directions for practice are but obscurely intimated; yet I have some reason not to disbelieve the process, without affirming or denying anything about the vertues of the remedy to be made by it. " Quando (saies he) oleum cinnamomi etc. suo sali alcali miscetur absque omni aqua, trium mensium artificiosa occultaque circulatione, totum in salem volatilem commutatum est, vere essentiam sui simplicis in nobis exprimit et usque in prima nostri constitutiva sese ingerit." A not unlike process he delivers in another place; from whence, if we suppose him to say true, I may argue, that since by the fire there may be produced a substance that is as well saline and volatile as the salt of hartshorn, blood, etc. which pass for elementary; and since that this volatile salt is really compounded of a chymical oyle and a fixt salt, the one made volatile by the other, and both associated by the fire, it may well be suspected that other substances, emerging upon the dissipation of bodies by the fire, may be new sorts of mixts, and consist of substances of differing natures; and particularly, I have sometimes suspected, that since the volatile salts of blood, hartshorn, etc. are fugitive and endowed with an exceeding strong smell, either that chymists do erroneously ascribe all odours to sulphurs, or that such salts consist of some oyly parts well incorporated with the saline ones. And the like conjecture I have also made concerning spirit of vinegar, which, though the chymists think one of the principles of that body, and though being an acid spirit it seems to be much less of kin than volatile salts to sulphurs; yet, not to mention its piercing smell; which I know not with what congruity the chymist will deduce from salt, I wonder they have not taken notice of what their own *Tyrocinium Chymicum* teach us concerning the distillation of *saccharum saturni ;* out of which Beguinus assures us, that he distilled, besides a very fine spirit, no less than two oyles, the one blood-red and ponderous, but the other swimming upon the top of the spirit, and of a yellow colour; of which he saies that he kept then some by him, to verify what he delivers. And though I remember not that I have had two distinct oyles from sugar of lead, yet that it will though distilled without addition yeeld some oyle, disagrees not with my experience. I know the chymists will be apt to pretend, that

these oyles are but the volatilized sulphur of lead; and
will perhaps argue it from what Beguinus relates, that
when the distillation is ended, you'l find a *caput mortuum*
extreamly black, and (as he speaks) *nullius momenti*, as
if the body, or at least the chief part of the metal itself
were by the distillation carried over the helme. But
since you know as well as I that *saccharum saturni* is a
kind of magistery, made only by calcining of lead *per se*,
dissolving it in distilled vinegar, and chrystalyzing the
solution; if I had leasure to tell you how differing a thing
I did upon examination find the *caput mortuum*, so slighted
by Beguinus, to be from what he represents it, I believe
you would think the conjecture proposed less probable
than one or other of these three; either that this oyle did
formerly concurr to constitute the spirit of vinegar, and
so that what passes for a chymical principle may yet be
further resoluble into distinct substances; or that some
parts of the spirit together with some parts of the lead
may constitute a chymical oyle, which therefore though
it pass for homogeneous, may be a very compounded
body: or at least that by the action of the distilled vinegar
and the saturnine calx one upon another, part of the
liquor may be so altered as to be transmuted from an
acid spirit into an oyle. And though the truth of either
of the two former conjectures would make the example
I have reflected on more pertinent to my present argument; yet you'l easily discern, the third and last conjecture cannot be unserviceable to confirm some other
passages of my discourse.

To return then to what I was saying just before
I mentioned Helmont's experiment, I shall subjoyne,
that chymists must confess also that in the perfectly
dephlegmed spirit of wine, or other fermented liquors,
that which they call the sulphur of the concrete loses, by
the fermentation, the property of oyle, (which the chymists
likewise take to be the true sulphur of the mixt) of being
unminglable with the water. And if you will credit
Helmont, a pound of the purest spirit of wine may barely
by the help of pure salt of tartar (which is but the fixed
salt of wine) be resolved or transmuted into scarce half

an ounce of salt, and as much elementary water as amounts
to the remaining part of the mentioned weight. And it
may (as I think I formerly also noted) be doubted,
whether that fixt and alcalizate salt, which is so unanimously agreed on to be the saline principle of incinerated
bodies, be not, as 'tis alcalizate, a production of the fire?
For though the taste of tartar, for example, seem to
argue that it contains a salt before it be burned, yet that
salt being very acid is of a quite differing taste from the
lixiviate salt of calcined tartar. And though it be not
truly objected against the chymists, that they obtain all
salts they make, by reducing the body they work on into
ashes with violent fires, (since hartshorn, amber, blood,
and divers other mixts yeeld a copious salt before they
be burned to ashes) yet this volatile salt differs much,
as we shall see anon, from the fixt alcalizate salt I speak
of; which for ought I remember is not producible by any
known way, without incineration. 'Tis not unknown to
chymists, that quicksilver may be precipitated, without
addition, into a dry powder, that remains so in water.
And some eminent spagyrists, and even Raimund Lully
himself, teach, that merely by the fire quicksilver may
in convenient vessels be reduced (at least in great part)
into a thin liquor like water, and minglable with it. So
that by the bare action of the fire, 'tis possible, that the
parts of a mixt body should be so disposed after new
and differing manners, that it may be sometimes of one
consistence, sometimes of another; and may in one state
be disposed to be mingled with water, and in another not.
I could also shew you, that bodies from which apart
chymists cannot obtain anything that is combustible,
may by being associated together, and by the help of the
fire, afford an inflamable substance. And that on the
other side, 'tis possible for a body to be inflamable, from
which it would very much puzzle any ordinary chymist,
and perhaps any other, to separate an inflamable principle
or ingredient. Wherefore, since the principles of chymists
may receive their denominations from qualities, which
it often exceeds not the power of art, nor alwaies that of
the fire to produce; and since such qualities may be

found in bodies that differ so much in other qualities from one another, that they need not be allowed to agree in that pure and simple nature, which principles, to be so indeed, must have; it may justly be suspected, that many productions of the fire that are shewed us by chymists, as the principles of the concrete that afforded them, may be but a new kind of mixts. And to annex, on this occasion, to these arguments taken from the nature of the thing, one of those which logicians call *ad hominem*, I shall desire you to take notice, that though Paracelsus himself, and some that are so mistaken as to think he could not be so, have ventured to teach, that not only the bodies here below, but the elements themselves, and all the other parts of the universe, are composed of salt, sulphur and mercury; yet the learned Sennertus, and all the more wary chymists, have rejected that conceit, and do many of them confess, that the *tria prima* are each of them made up of the four elements; and others of them make earth and water concurr with salt, sulphur and mercury, to the constitution of mixt bodies. So that one sort of these spagyrists, notwithstanding the specious titles they give to the productions of the fire, do in effect grant what I contend for. And, of the other sort I may well demand, to what kind of bodies the phlegm and dead earth, to be met with in chymical resolutions, are to be referred? For either they must say, with Paracelsus, but against their own concessions, as well as against experience, that these are also composed of the *tria prima*, whereof they cannot separate any one from either of them; or else they must confess that two of the vastest bodies here below, earth and water, are neither of them composed of the *tria prima ;* and that consequently those three are not the universal and adequate ingredients, neither of all sublunary bodies, nor even of all mixt bodies.

I know that the chief of these chymists represent, that though the distinct substances into which they divide mixt bodies by the fire, are not pure and homogeneous; yet since the four elements into which the Aristotelians pretend to resolve the like bodies by the same agent, are not simple neither, as themselves acknowledge, 'tis as allowable for the chymists to call the one principles, as for the peripateticks to call the other elements, since in both cases the imposition of the name is grounded only upon the predominancy of that element whose name is ascribed to it. Nor shall I deny, that this argument of the chymists is no ill one against the Aristotelians. But what answer can it prove to me, who you know am disputing as well against the Aristotelian elements, as the chymical principles, and must not look upon any body as a true principle or element, but as yet compounded, which is not perfectly homogeneous, but is further resoluble into any number of distinct substances how small soever. And as for the chymists calling a body salt, or sulphur, or mercury, upon pretence that the principle of the same name is predominant in it, that itself is an acknowledgment of what I contend for; namely that these productions of the fire are yet compounded bodies. And yet whilst this is granted, it is affirmed, but not proved, that the reputed salt, or sulphur, or mercury, consists mainly of one body that deserves the name of a principle of the same denomination. For how do chymists make it appear that there are any such primitive and simple bodies in those we are speaking of; since 'tis upon the matter confessed by the answer lately made, that these are not such? And if they pretend by reason to evince what they affirm, what becomes of their confident boasts, that the chymist (whom they therefore, after Beguinus, call a *philosophus* or *opifex sensatus*) can convince our eyes, by manifestly shewing in any mixt body those simple substances he teaches them to be composed of? And indeed, for the chymists to have recourse in this case to other proofs than experiments, as it is to wave the grand argument that has all this while been given out for a demonstrative one; so it releases me from the obligation to prosecute a dispute wherein I am not engaged to examine any but experimental proofs. I know it may plausibly enough be represented, in favour of the chymists, that it being evident that much the greater part of anything they call salt, or sulphur, or mercury, is really such;

it would be very rigid to deny those substances the names ascribed them, only because of some slight mixture of another body; since not only the peripateticks call particular parcels of matter elementary, though they acknowledge that elements are not to be anywhere found pure, at least here below; and since especially there is a manifest analogie and resemblance betwixt the bodies obtainable by chymical anatomies and the principles whose names are given them; I have, I say, considered that these things may be represented; but as for what is drawn from the custome of the peripateticks, I have already told you, that though it may be employed against them, yet it is not available against me, who allow nothing to be an element that is not perfectly homogeneous. And whereas it is alledged, that the predominant principle ought to give a name to the substance wherein it abounds; I answer, that that might much more reasonably be said, if either we or the chymists had seen nature take pure salt, pure sulphur, and pure mercury, and compound of them every sort of mixt bodies. But, since 'tis to experience that they appeal, we must not take it for granted, that the distilled oyle (for instance) of a plant is mainly composed of the pure principle called sulphur, till they have given us an ocular proof, that there is in that sort of plants such an homogeneous sulphur. For as for the specious argument, which is drawn from the resemblance betwixt the productions of the fire, and the respective, either Aristotelian elements, or chymical principles, by whose names they are called; it will appear more plausible than cogent, if you will but recall to mind the state of the controversie; which is not, whether or no there be obtained from mixt bodies certain substances that agree in outward appearance, or in some qualities with quicksilver or brimstone, or some such obvious or copious body; but whether or no all bodies confessed to be perfectly mixt were composed of, and are resoluble into a determinate number of primary unmixt bodies. For, if you keep the state of the question in your eye, you'l easily discerne that there is much of what should be demonstrated, left unproved by those chymical experiments we are examin-

ing. But (not to repeat what I have already discovered
more at large) I shall now take notice, that it will not
presently follow, that because a production of the fire has
some affinity with some of the greater masses of matter
here below, that therefore they are both of the same nature,
and deserve the same name; for the chymists are not
content, that flame should be lookt upon as a parcel of the
element of fire, though it be hot, dry, and active, because
it wants some other qualities belonging to the nature of
elementary fire. Nor will they let the peripateticks call
ashes, or quicklime, earth, notwithstanding the many
likenesses between them; because they are not tasteless, as
elementary earth ought to be: but if you should ask me,
what then it is, that all the chymical anatomies of bodies
do prove, if they prove not that they consist of the three
principles into which the fire resolves them? I answer
that their dissections may be granted to prove, that some
mixt bodies (for in many it will not hold) are by the fire,
when they are included in close vessels, (for that condition
also is often requisite) dissoluble into several substances
differing in some qualities, but principally in consistence.
So that out of most of them may be obtained a fixt
substance partly saline, and partly insipid, an unctuous
liquor, and another liquor or more that without being
unctuous have a manifest taste. Now if chymists will
agree to call the dry and sapid substance salt, the unctuous
liquor sulphur, and the other mercury, I shall not much
quarrel with them for so doing: but if they will tell me
that salt, sulphur, and mercury, are simple and primary
bodies whereof each mixt body was actually compounded,
and which was really in it antecedently to the operation
of the fire, they must give me leave to doubt whether
(whatever their other arguments may do) their experiments prove all this. And if they will also tell me that
the substances their anatomies are wont to afford them,
are pure and similar, as principles ought to be, they must
give me leave to believe my own senses; and their own
confessions, before their bare assertions. And that you
may not (Eleutherius) think I deal so rigidly with them,
because I scruple to take these productions of the fire for

such as the chymists would have them pass for, upon the
account of their having some affinity with them; consider
a little with me, that in regard an element or principle
ought to be perfectly similar and homogeneous, there is
no just cause why I should rather give the body proposed
the name of this or that element or principle, because it
has a resemblance to it in some obvious quality, rather
than deny it that name upon the account of divers other
qualities, wherein the proposed bodies are unlike; and if
you do but consider what slight and easily producible
qualities they are that suffice, as I have already more than
once observed, to denominate a chymical principle or
an element, you'l not, I hope, think my wariness to be
destitute either of example, or else of reason. For we
see that the chymists will not allow the Aristotelians that
the salt in ashes ought to be called earth, though the saline
and terrestrial part symbolize in weight, in dryness, in
fixness and fusibility, only because the one is sapid and
dissoluble in water, and the other not: besides, we see
that sapidness and volatility are wont to denominate the
chymists mercury or spirit; and yet how many bodies,
think you, may agree in those qualities which may yet be
of very differing natures, and disagree in qualities either
more numerous, or more considerable, or both. For not
only spirit of nitre, *aqua fortis*, spirit of salt, spirit of oyle
of vitriol, spirit of allume, spirit of vinegar, and all saline
liquors distilled from animal bodies, but all the acetous
spirits of woods freed from their vinegar; all these, I
say, and many others must belong to the chymists
mercury, though it appear not why some of them should
more be comprehended under one denomination than the
chymists sulphur, or oyle should likewise be; for their
distilled oyles are also fluid, volatile, and tastable, as well
as their mercury; nor is it necessary, that their sulphur
should be unctuous or dissoluble in water, since they
generally referr spirit of wine to sulphurs, although that
spirit be not unctuous, and will freely mingle with water.
So that bare inflamability must constitute the essence
of the chymists sulphur; as uninflamableness joyned
with any taste is enough to intitle a distilled liquor to be

their mercury. Now since I can further observe to you, that spirit of nitre and spirit of hartshorne being poured together will boyle and hisse and tosse up one another into the air, which the chymists make signes of great antipathy in the natures of bodies, (as indeed these spirits differ much both in taste, smell, and operations) since I elsewhere tell you of my having made two sorts of oyle out of the same man's blood, that would not mingle with one another; and since I might tell you divers examples I have met with, of the contrariety of bodies which according to the chymists must be huddled up together under one denomination; I leave you to judge whether such a multitude of substances as may agree in these slight qualities, and yet disagree in others more considerable, are more worthy to be called by the name of a principle (which ought to be pure and homogeneous) than to have appellations given them that may make them differ, in name too, from the bodies from which they so wildly differ in nature. And hence also, by the by, you may perceive that 'tis not unreasonable to distrust the chymists way of argumentation, when being unable to shew us that such a liquor is (for example) purely saline, they prove, that at least salt is much the predominant principle, because that the proposed substance is strongly tasted, and all taste proceeds from salt; whereas those spirits, such as spirit of tartar, spirit of hartshorn, and the like, which are reckoned to be the mercuries of the bodies that afford them, have manifestly a strong and piercing taste, and so has (according to what (I formerly noted the spirit of box, etc. even after the acid liquor that concurred to compose it has been separated from it. And indeed, if sapidness belong not to the spirit or mercurial principle of vegetables and animals: I scarce know how it will be discriminated from their phlegm, since by the absence of inflamability it must be distinguished from their sulphur which affords me another example, to prove how unacurate the chymical doctrine is in our present case; since not only the spirits of vegetables and animals, but their oyles are very strongly tasted, as he that shall but wet his tongue with chymical oyle of cinnamon or of cloves, or even of turpentine, may quickly find, to his smart. And not only I never tryed any chymical oyles whose taste was not very manifest and strong; but a skilful and inquisitive person who made it his business by elaborate operations to depurate chymical oyles, and reduce them to an elementary simplicity, informes us, that he never was able to make them at all tasteless; whence I might inferr, that the proof chymists confidently give us of a bodies being saline, is so far from demonstrating the predominancy, that it does not clearly evince so much as the presence of the saline principle in it. But I will not (pursues Carneades) remind you, that the volatile salt of hartshorn, amber, blood, etc. are exceeding strongly scented, notwithstanding that most chymists deduce odours from sulphur, and from them argue the predominancy of that principle in the odorous body, because I must not so much as add any new examples of the incompetency of this sort of chymical arguments; since having already detained you but too long in those generals that appertain to my fourth consideration 'tis time that I proceed to the particulars themselves, to which I thought fit they should be previous.

These generals (continues Carneades) being thus premised, we might the better survey the unlikeness that an attentive and unprepossessed observer may take notice of in each sort of bodies which the chymists are wont to call the salts or sulphurs or mercuries of the concretes that yeeld them, as if they had all a simplicity, and identity of nature: whereas salts if they were all elementary would as little differ as do the drops of pure and simple water. 'Tis known that both chymists and physitians ascribe to the fixt salts of calcined bodies the vertues of their concretes; and consequently very differing operations. So we find the alcali of wormwood much commended in distempers of the stomach; that of eyebright for those that have a weak sight; and that of guajacum (of which a great quantity yeelds but a very little salt) is not only much commended in venereal diseases, but is believed to have a peculiar purgative vertue, which yet

I have not had occasion to try. And though, I confess,
I have long thought, that these alcalizate salts are, for
the most part, very near of kin, and retain very little
of the properties of the concretes whence they were
separated; yet being minded to observe watchfully
whether I could meet with any exceptions to this general
observation, I observed at the glass-house, that sometimes the metal (as the workmen call it) or mass of colliquated ingredients, which by blowing they fashion into
vessels of divers shapes, did sometimes prove of a very
differing colour, and a somewhat differing texture, from
what was usual. And having enquired whether the
cause of such accidents might not be derived from the
peculiar nature of the fixt salt employed to bring the sand
to fusion, I found that the knowingst workmen imputed
these misadventures to the ashes, of some certain kind
of wood, as having observed the ignobler kind of glass
I lately mentioned to be frequently produced, when they
had employed such sorts of ashes, which therefore they
scruple to make use of, if they took notice of them beforehand. I remember also, that an industrious man of my
acquaintance having bought a vast quantity of tobacco
stalks to make a fixt salt with, I had the curiosity to
go see whether that exotick plant, which so much
abounds in volatile salt, would afford a peculiar kind of
alcali; and I was pleased to find that in the lixivium of
it, it was not necessary, as is usual, to evaporate all the
liquor, that there might be obtained a saline calx, consisting like lime quenched in the air of a heap of little corpuscles of unregarded shapes: but the fixt salt shot into
figured chrystal, almost as nitre or sal armoniack and
other uncalcined salts are wont to do; and I further
remember that I have observed that in the fixt salt of
urine, brought by depuration to be very white, a taste not
so unlike to that of common salt, and very differing from
the wonted caustick lixiviate taste of other salts made by
incineration. But because the instances I have alledged
of the difference of alcalizate salt are but few, and therefore I am still inclined to think, that most chymists and
many physitians do, inconsiderately enough and without

warrant from experience, ascribe the vertues of the concretes exposed to calcination, to the salts obtained by it; I shall rather to shew the disparity of salts mention in the first place the apparent difference betwixt the vegetable fixt salts and the animal volatile ones: as (for example) betwixt salt of tartar, and salt of hartshorn; whereof the former is so fixt that 'twill indure the brunt of a violent fire, and stand in fusion like a metal; whereas the other (besides that it has a differing taste and a very differing smell) is so far from being fixt, that it will fly away in a gentle heat as easily as spirit of wine itself. And to this I shall add, in the next place, that even among the volatile salts themselves, there is a considerable difference, as appears by the distinct properties of (for instance) salt of amber, salt of urine, salt of man's skull, (so much extolled against the falling sickness) and divers others which cannot escape an ordinary observer. And this diversity of volatile salts I have observed to be sometimes discernable even to the eye, in their figures. For the salt of hartshorn I have observed to adhere to the receiver in the forme almost of a parallelipipedon; and of the volatile salt of humane blood (long digested before distillation, with spirit of wine) I can shew you store of grains of that figure which geometricians call a rhombus; though I dare not undertake that the figures of these or other saline chrystals (if I may so call them) will be alwaies the same, whatever degree of fire have been employed to force them up, or how hastily soever they have been made to convene in the spirits or liquors, in the lower part of which I have usually observed them after a while to shoot. And although, as I lately told you, I seldom found any difference, as to medical vertues, in the fixt salts of divers vegetables; and accordingly I have suspected that most of these volatile salts, having so great a resemblance in smell, in taste, and fugitiveness, differ but little, if at all, in their medicinal properties: as indeed I have found them generally to agree in divers of them (as in their being somewhat diaphoretick and very deopilative) yet I remember Helmont somewhere informs us, that there is this difference betwixt the saline spirit of

urine and that of man's blood, that the former will not
cure the epilepsy, but the latter will. Of the efficacy
also of the salt of common amber against the same disease
in children, (for in grown persons it is not a specifick) I may
elsewhere have an occasion to entertain you. And when
I consider that to the obtaining of these volatile salts
(especially that of urine) there is not requisite such a
destructive violence of the fire, as there is to get those
salts that must be made by incineration, I am the more
invited to conclude, that they may differ from one another
and consequently recede from an elementary simplicity.
And, if I could here shew you what Mr. Boyle has observed,
touching the various chymical distinctions of salts; you
would quickly discern, not only that chymists do give
themselves a strange liberty to call concretes salts, that
are according to their own rules to be looked upon as
very compounded bodies; but that among those very
salts that seem elementary, because produced upon the
anatomy of the bodies that yeeld them, there is not only
a visible disparity, but, to speak in the common language,
a manifest antipathy or contrariety: as is evident in the
ebullition and hissing that is wont to ensue, when the acid
spirit of vitriol, for instance, is poured upon hot ashes, or
salt of tartar. And I shall beg leave of this gentleman,
(saies Carneades) casting his eyes on me, to let me observe
to you out of some of his papers, particularly those wherein
he treats of some preparations of urine, that not only one
and the same body may have two salts of a contrary
nature, as he exemplifies in the spirit and alkali of nitre;
but that from the same body there may without addition
be obtained three differing and visible salts. For he
relates, that he observed in urine, not only a volatile and
chrystalline salt, and a fixt salt, but likewise a kind of
sal armoniack, or such a salt as would sublime in the form
of a salt, and therefore was not fixt, and yet was far from
being so fugitive as the volatile salt; from which it seemed
also otherwise to differ. I have indeed suspected that
this may be a sal armoniack properly enough so called, as
compounded of the volatile salt of urine, and the fixt
of the same liquor, which, as I noted, is not unlike sea-

salt; but that itself argues a manifest difference betwixt the salts, since such a volatile salt is not wont to unite thus with an ordinary alcali, but to fly away from it in the heat. And on this occasion I remember, that to give some of my friends an ocular proof of the difference betwixt the fixt and volatile salt of (the same concrete) wood, I devised the following experiment. I took common Venetian sublimate, and dissolved as much of it as I well could in fair water: then I took wood ashes, and pouring on them warme water, dissolved their salt; and filtrating the water, as soon as I found the lixivium sufficiently sharp upon the tongue, I reserved it for use: then one part of the former solution of sublimate dropping a little of this dissolved fixt salt of wood, the liquors presently turned of an orange colour; but upon the other part of the clear solution of sublimate putting some of the volatile salt of wood (which abounds in the spirit of soot) the liquor immediately turned white, almost like milke, and after a while let fall a white sediment, as the other liquor did a yellow one. To all this that I have said concerning the difference of salts, I might add what I formerly told you, concerning the simple spirit of box, and such like woods, which differ much from the other salts hitherto mentioned, and yet would belong to the saline principle, if chymists did truly teach that all tastes proceed from it. And I might also annex, what I noted to you out of Helmont concerning bodies, which, though they consist in great part of chymical oyles, do yet appear but volatile salts; but to insist on these things, were to repeat; and therefore I shall proceed.

This disparity is also highly eminent in the separated sulphurs or chymical oyles of things. For they contain so much of the scent, and taste, and vertues, of the bodies whence they were drawn, that they seem to be but the material *crasis* (if I may so speak) of their concretes. Thus the oyles of cinnamon, cloves, nutmegs and other spices, seem to be but the united aromatick parts that did ennoble those bodies. And 'tis a known thing, that oyl of cinnamon, and oyle of cloves, (which I have likewise observed in the oyles of several woods) will sink to the bottom of water: whereas those of nutmegs and divers other vegetables will swim upon it. The oyle (abusively called spirit) of roses swims at the top of the water in the forme of a white butter, which I remember not to have observed in any other oyle drawn in any limbeck; yet there is a way (not here to be declared) by which I have seen it come over in the forme of other aromatick oyles, to the delight and wonder of those that beheld it. In oyle of aniseeds, which I drew both with, and without fermentation, I observed the whole body of the oyle in a cool place to thicken into the consistence and appearance of white butter, which with the least heat resumed its former liquidness. In the oyle of olive drawn over in a retort, I have likewise more than once seen a spontaneous coagulation in the receiver: and I have of it by me thus congealed; which is of such a strangely penetrating scent, as if 'twould perforate the noses that approach it. The like pungent odour I also observed in the distilled liquor of common sope, which forced over from minium, lately afforded an oyle of a most admirable penetrancy; and he must be a great stranger, both to the writings and preparations of chymists, that sees not in the oyles they distill from vegetables and animals, a considerable and obvious difference. Nay I shall venture to add, Eleutherius (what perhaps you will think of kin to a paradox) that divers times out of the same animal or vegetable, there may be extracted oyles of natures obviously differing. To which purpose I shall not insist on the swimming and sinking oyles, which I have sometimes observed to float on, and subside under the spirit of guajacum, and that of divers other vegetables distilled with a strong and lasting fire; nor shall I insist on the observation elsewhere mentioned, of the divers and unmingleable oyles afforded us by humane blood long fermented and digested with spirit of wine, because these kind of oyles may seem chiefly to differ in consistence and weight, being all of them high coloured and adust. But the experiment, which I devised to make out this difference of the oyles of the same vegetable, *ad oculum,* (as they speak) was this that followes. I took a pound of aniseeds, and having grosly beaten them, caused them to be put into a very large glass retort almost filled with fair water; and placing this retort in a sand furnace, I caused a very gentle heat to be administred during the first day, and a great part of the second, till the water was for the most part drawn off, and had brought over with it at least most of the volatile and aromatick oyle of the seeds. And then encreasing the fire, and changing the receiver, I obtained besides an empyreumatical spirit, a quantity of adust oyle; whereof a little floated upon the spirit, and the rest was more heavy, and not easily separable from it. And whereas these oyles were very dark, and smelled (as chymists speak) so strongly of the fire, that their odour did not betray from what vegetables they had been forced; the other *aromatick* oyle was enriched with the genuine smell and taste of the concrete; and spontaneously coagulating itself into white butter did manifest itself to be the true oyle of aniseeds; which concrete I therefore chose to employ about this experiment, that the difference of these oyles might be more conspicuous than it would have been, had I instead of it destilled another vegetable.

I had almost forgot to take notice, that there is another sort of bodies, which though not obtained from concretes by distillation, many chymists are wont to call their sulphur; not only because such substances are, for the most part, high coloured, (whence they are also, and that more properly, called tinctures) as dissolved sulphurs are wont to be; but especially because they are, for the most part, abstracted and separated from the rest of the mass by spirit of wine: which liquor those men supposing to be sulphureous, they conclude, that what it works upon, and abstracts, must be a sulphur also. And upon this account they presume, that they can sequester the sulphur even of minerals and metalls; from which 'tis known that they cannot by fire alone separate it. To all this I shall answer; That if these sequestred substances were indeed the sulphurs of the bodies whence they are drawn, there would as well be a great disparity betwixt chymical sulphurs obtained by spirit of wine, as I have already shewn there is betwixt those obtained by distillation in the forme of oyles: which will be evident from hence, that not to urge that themselves ascribe distinct vertues to mineral tincture, extolling the tincture of gold against such and such diseases; the tincture of antimony, or of its glass, against others; and the tincture of emerald against others; 'tis plain, that in tinctures drawn from vegetables, if the superfluous spirit of wine be distilled off, it leaves at the bottom that thicker substance which chymists use to call the extract of the vegetable. And that these extracts are endowed with very differing qualities according to the nature of the particular bodies that afforded them (though I fear seldom with so much of the specifick vertues as is wont to be imagined) is freely confessed both by physitians and chymists. But Eleutherius (saies Carneades) we may here take notice that the chymists do as well in this case, as in many others allow themselves a license to abuse words: for not again to argue from the differing properties of tinctures, that they are not exactly pure and elementary sulphurs; they would easily appear not to be so much as sulphur's, although we should allow chymical oyles to deserve that name. For however in some mineral tinctures the natural fixtness of the extracted body does not alwaies suffer it to be easily further resoluble into differing substances; yet in very many extracts drawn from vegetables, it may very easily be manifested that the spirit of wine has not sequestred the sulphureous ingredient from the saline and mercurial ones; but has dissolved (for I take it to be a solution) the finer parts of the concrete (without making any nice distinction of their being perfectly sulphureous or not) and united itself with them into a kind of magistery which consequently must contain ingredients or parts of several sorts. For we see that the stones that are rich in vitriol, being often drenched with rain-water, the liquor will then extract a fine and transparent substance coagulable into vitriol; and yet though this vitriol be readily dissoluble in water, it is not a true elementary salt, but, as you know, a body resoluble into very differing parts, whereof one (as I shall have occasion to tell you anon) is yet of a metalline, and consequently not of an elementary nature. You may consider also, that common sulphur is readily dissoluble in oyle of turpentine, though notwithstanding its name it abounds as well, if not as much, in salt as in true sulphur; witness the great quantity of saline liquor it affords being set to flame away under a glass bell. Nay I have, which perhaps you will think strange, with the same oyle of turpentine alone easily enough dissolved crude antimony finely powdered into a blood-red balsam, wherewith perhaps considerable things may be performed in surgery. And if it were now requisite, I could tell you of some other bodies, (such as perhaps you would not suspect) that I have been able to work upon with certain chymical oyles. But instead of digressing further I shall make this use of the example I have named. That 'tis not unlikely, but that spirit of wine which by its pungent taste, and by some other qualities that argue it better, (especially its reducib eness, according to Helmont, into alcali and water), seems to be as well of a saline as of a sulphureous nature, may well be supposed capable of dissolving substances that are not merely elementary sulphurs, though perhaps they may abound with parts that are of kin thereunto. For I find that spirit of wine will dissolve *gumm lacca, benzoine,* and the *resinous* parts of *jallap,* and even of *guajacum;* whence we may well suspect that it may from spices, herbs, and other less compacted vegetables, extract substances that are not perfect sulphurs but mixt bodies. And to put it past dispute, there is many a vulgar extract drawn with spirit of wine, which committed to distillation will afford such differing substances as will loudly proclaim it to have been a very compounded body. So that we may justly suspect, that even in mineral tinctures it will not alwaies follow, that because a red substance is drawn from the concrete by spirit of wine, that substance is its true and elementary sulphur. And though some of these extracts may perhaps be inflamable; yet, besides that others are not, and besides that their being reduced to such minuteness of parts may much facilitate their taking fire; besides this, I say, we see that common sulphur, common oyle, gumm lac,

and many unctuous and resinous bodies, will flame well
enough, though they be of very compounded natures:
nay travellers of unsuspected credit assure us, as a known
thing, that in some northern countries where firr trees and
pines abound, the poorer sort of inhabitants use long
splinters of those resinous woods to burn instead of
candles. And as for the redness wont to be met with in
such solutions, I could easily shew, that 'tis not necessary
it should proceed from the sulphur of the concrete, dissolved by the spirit of wine; if I had leasure to manifest
how much chymists are wont to delude themselves and
others, by the ignorance of those other causes, upon whose
account spirit of wine and other menstruum may acquire
a red or some other high colour. But to returne to our
chymical oyles, supposing that they were exactly pure; yet
I hope they would be, as the best spirit of wine is, but the
more inflamable and deflagrable. And therefore since
an oyle can be by the fire alone immediately turned into
flame, which is something of a very differing nature from
it: I shall demand how this oyle can be a primogeneal
and incorruptible body, as most chymists would have
their principles; since it is further resoluble into flame,
which whether or no it be a portion of the element of fire,
as an Aristotelian would conclude, is certainly something
of a very differing nature from a chymical oyle, since it
burnes, and shines, and mounts swiftly upwards; none
of which a chymical oyle does, whilst it continues such.
And if it should be objected, that the dissipated parts of
this flaming oyle may be caught and collected again into
oyl or sulphur; I shall demand, what chymist appears
to have ever done it; and without examining whether
it may not hence be as well said that sulphur is but compacted fire, as that fire is but diffused sulphur, I shall
leave you to consider whether it may not hence be argued,
that neither fire nor sulphur are primitive and indestructible bodies; and I shall further observe that at least
it will hence appear, that a portion of matter may, without
being compounded with new ingredients, by having the
texture and motion of its small parts changed, be easily,
by the means of the fire, endowed with new qualities, more

differing from them it had before, than are those which suffice to discriminate the chymists principles from one another.

We are next to consider, whether in the anatomy of mixt bodies, that which chymists call the mercurial part of them be uncompounded, or no. But to tell you true, though chymists do unanimously affirm that their resolutions discover a principle, which they call mercury, yet I find them to give of it descriptions so differing, and so ænigmatical, that I, who am not ashamed to confess that I cannot understand what is not sence, must acknowledge to you that I know not what to make of them. Paracelsus himself, and therefore, as you will easily believe, many of his followers, does somewhere call that mercury which ascends upon the burning of wood, as the peripateticks are wont to take the same smoake for air; and so seems to define mercury by volatility, or (if I may coyne such a word) effumability. But since, in this example, both volatile salt and sulphur make part of the smoake, which does indeed consist also both of phlegmatick and terrene corpuscles, this notion is not to be admitted; and I find that the more sober chymists themselves disavow it. Yet to shew you how little of clearness we are to expect in the accounts even of later spagyrists, be pleased to take notice, that Beguinus, even in his *Tyrocinium Chymicum,* written for the instruction of novices, when he comes to tell us what are meant by the *tria prima,* which for their being principles ought to be defined the more accurately and plainly, gives us this description of mercury; " Mercurius (saies he) est liquor ille acidus, permeabilis, penetrabilis, æthereus, ac purissimus, à quo omnis nutricatio, sensus, motus, vires, colores, senectutisque præproperæ retardatio." Which words are not so much a definition of it, as an encomium: and yet Quercetanus in his description of the same principle adds to these divers other epithets. But both of them, to skip very many other faults that may be found with their metaphorical descriptions, speak incongruously to the chymists own principles. For if mercury be an acid liquor, either hermetical philosophy must err in ascribing all tastes to salt, or else mercury must not be a principle, but compounded of a saline ingredient and somewhat else. Libavius, though he find great fault with the obscurity of what the chymists write concerning their mercurial principle, does yet but give us such a negative description of it, as Sennertus, how favourable soever to the *tria prima*, is not satisfied with. And this Sennertus himself, though the learnedest champion for the hypostatical principles, does almost as frequently as justly complain of the unsatisfactoriness of what the chymists teach concerning their mercury; and yet he himself (but with his wonted modesty) substitutes instead of the description of Libavius, another, which many readers, especially if they be not peripateticks, will not know what to make of. For scarce telling us any more, than that in all bodies that which is found besides salt and sulphur, and the elements, or, as they call them, phlegm and dead earth, is that spirit which in Aristotle's language may be called ὀυσία ἀναλογω τ῀ϛδ ἄστρων στοιχείῳ. He saies that which I confess is not at all satisfactory to me, who do not love to seem to acquiesce in any man's mystical doctrines, that I may be thought to understand them.

If (saies Eleutherius) I durst presume that the same thing would be thought clear by me, and those that are fond of such cloudy expressions as you justly tax the chymists for, I should venture to offer to consideration, whether or no, since the mercurial principle that arises from distillation is unanimously asserted to be distinct from the salt and sulphur of the same concrete, that may not be called the mercury of a body, which though it ascend in distillation, as do the phlegme and sulphur, is neither insipid like the former, nor inflamable like the latter. And therefore I would substitute to the too much abused name of mercury, the more clear and familiar appellation of spirit, which is also now very much made use of even by the chymists themselves of our times, though they have not given us so distinct an explication, as were fit, of what may be called the spirit of a mixt body.

I should not perhaps (saies Carneades) much quarrel

with your notion of mercury. But as for the chymists, what they can mean, with congruity to their own principles, by the mercury of animals and vegetables, 'twill not be so easie to find out; for they ascribe tastes only to the saline principle, and consequently would be much put to it to shew what liquor it is, in the resolution of bodies, that not being insipid, for that they call phlegme, neither is inflamable as oyle or sulphur, nor has any taste; which according to them must proceed from a mixture, at least, of salt. And if we should take spirit in the sence of the word received among modern chymists and physitians, for any distilled liquor that is neither phlegme nor oyle, the appellation would yet appear ambiguous enough. For plainly, that which first ascends in the distillation of wine and fermented liquors, is generally as well by chymists as others reputed a spirit. And yet pure spirit of wine being wholly inflamable ought according to them to be reckoned to the sulphureous, not the mercurial principle. And among the other liquors that go under the name of spirits, there are divers which seem to belong to the family of salts, such as are the spirits of nitre, vitriol, sea-salt and others, and even the spirit of hartshorn, being, as I have tryed, in great part, if not totally reducible into salt and phlegme, may be suspected to be but a volatile salt disguised by the phlegme mingled with it into the forme of a liquor. However if this be a spirit, it manifestly differs very much from that of vinegar, the taste of the one being acid, and the other salt, and their mixture in case they be very pure, sometimes occasioning an effervescence like that of those liquors the chymists count most contrary to one another. And even among those liquors that seem to have a better title, than those hitherto mentioned, to the name of spirits, there appears a sensible diversity; for spirit of oak, for instance, differs from that of tartar, and this from that of box, or of guajacum. And in short, even these spirits as well as other distilled liquors manifest a great disparity betwixt themselves, either in their actions on our senses, or in their other operations.

And (continues Carneades) besides this disparity that

is to be met with among those liquors that the moderns call spirits, and take for similar bodies, what I have formerly told you concerning the spirit of boxwood may let you see that some of those liquors not only have qualities very differing from others, but may be further resolved into substances differing from one another.

And since many moderne chymists and other naturalists are pleased to take the mercurial spirit of bodies for the same principle, under differing names, I must invite you to observe, with me, the great difference that is conspicuous betwixt all the vegetable and animal spirits I have mentioned and running mercury. I speak not of that which is commonly sold in shops that many of themselves will confesse to be a mixt body; but of that which is separated from metals, which by some chymists that seem more philosophers than the rest, and especially by the above mentioned Claveus, is (for distinction sake) called *mercurius corporum*. Now this metalline liquor being one of those three principles of which mineral bodies are by spagyrists affirmed to be composed and to be resoluble into them, the many notorious differences betwixt them and the mercuries, as they call them, of vegetables and animals will allow me to inferr, either that minerals and the other two sorts of mixt bodies consist not of the same elements, or that those principles whereinto minerals are immediately resolved, which chymists with great ostentation shew us as the true principles of them, are but secondary principles, or mixts of a peculiar sort, which must be themselves reduced to a very differing forme, to be of the same kind with vegetable and animal liquors.

But this is not all; for although I formerly told you how little credit there is to be given to the chymical processes commonly to be met with, of extracting the mercuries of metals, yet I will now add, that supposing that the more judicious of them do not untruly affirme that they have really drawn true and running mercury from several metals (which I wish they had clearly taught us how to do also,) yet it may be still doubted whether such extracted mercuries do not as well differ from common quicksilver, and from one another, as from the mercuries of vegetables and animals. Claveus, in his Apology, speaking of some experiments whereby metalline mercuries may be fixt into the nobler metals, adds, that he spake of the mercuries drawn from metals; because common quicksilver by reason of its excessive coldness and moisture is unfit for that particular kind of operation; for which though a few lines before he prescribes in general the mercuries of metalline bodies, yet he chiefly commends that drawn by art from silver. And elsewhere, in the same book, he tells us, that he himself tryed, that by bare coction the quicksilver of tin or pewter (*argentum vivum ex stanno prolicitum*) may by an efficient cause, (as he speaks) be turned into pure gold. And the experienced Alexander van Suchten, somewhere tells us, that by a way he intimates may be made a mercury of copper, not of the silver colour of other mercuries, but green; to which I shall add, that an eminent person, whose name his travells and learned writings have made famous, lately assured me that he had more than once seen the mercury of lead (which whatever authors promise, you will find it very difficult to make, at least in any considerable quantity) fixt into perfect gold. And being by me demanded whether or no any other mercury would not as well have been changed by the same operations, he assured me of the negative.

And since I am fallen upon the mention of the mercuries of metals, you will perhaps expect, (Eleutherius) that I should say something of their two other principles; but I must freely confess to you, that what disparity there may be between the salts and sulphurs of metals or other minerals, I am not myself experienced enough in the separations and examens of them, to venture to determine: (for as for the salts of metals, I formerly represented it as a thing much to be questioned, whether they have any at all.) And for the processes of separation I find in authors, if they were (what many of them are not) successfully practicable, as I noted above, yet they are to be performed by the assistance of other bodies, so hardly, if upon any termes at all, separable from them, that it is very difficult to give the separated principles all their due, and no more. But the sulphur of antimony which is vehemently vomitive, and the strongly scented anodyne sulphur of vitriol inclines me to think that not only mineral sulphurs differ from vegetable ones, but also from one another, retaining much of the nature of their concretes. The salts of metals, and of some sort of minerals, you will easily guesse (by the doubts I formerly expressed, whether metals have any salt at all) that I have not been so happy as yet to see, perhaps not for want of curiosity. But if Paracelsus did alwaies write so consentaneously to himself that his opinion were confidently to be collected from every place of his writings where he seems to expresse it, I might safely take upon me to tell you, that he both countenances in general what I have delivered in my fourth main consideration, and in particular warrants me to suspect that there may be a difference in metalline and mineral salts, as well as we find it in those of other bodies. For, " Sulphur (saies he) aliud in auro, aliud in argento, aliud in ferro, aliud in plumbo, stanno, etc. sic aliud in saphyro, aliud in smaragdo, aliud in rubino, chrysolitho, amethysto, magnete, etc. Item aliud in lapidibus, silice, salibus, fontibus, etc. nec vero tot sulphura tantum, sed et totidem salia; sal aliud in metallis, aliud in gemmis, aliud in lapidibus, aliud in salibus, aliud in vitriolo, aliud in alumine: similis etiam mercurii est ratio. Alius in metallis, alius in gemmis, etc. Ita ut unicuique speciei suus peculiaris mercurius sit. Et tamen res saltem tres sunt; una essentia est sulphur; una est sal; una est mercurius. Addo quod et specialius adhuc singula dividantur; aurum enim non unum, sed multiplex, ut et non unum pyrum, pomum, sed idem multiplex, totidem etiam sulphura auri, salia auri, mercurii auri; idem competit etiam metallis et gemmis; ut quot saphyri præstantiores, læviores, etc. tot etiam saphyrica sulphura, saphyrica salia, saphyrici mercurii, etc. Idem verum etiam est de turconibus et gemmis aliis universis." From which passage (Eleutherius) I suppose you will think I might without rashness conclude, either that my opinion is favoured by that of Paracelsus, or that

Paracelsus his opinion was not alwaies the same. But because in divers other places of his writings he seems to talk at a differing rate of the three principles and the four elements, I shall content myself to inferr from the alledged passage, that if his doctrine be not consistent with that part of mine which it is brought to countenance, it is very difficult to know what his opinion concerning salt, sulphur and mercury, was; and that consequently we had reason about the beginning of our conferences, to decline taking upon us, either to examine or oppose it.

I know not whether I should on this occasion add, that those very bodies, the chymists call phlegme and earth, do yet recede from an elementary simplicity. That common earth and water frequently do so, notwithstanding the received contrary opinion, is not denyed by the more wary of the moderne peripateticks themselves: and certainly most earths are much less simple bodies than is commonly imagined even by chymists, who do not so considerately to prescribe and employ earths promiscuously in those distillations that require the mixture of some *caput mortuum*, to hinder the flowing together of the matter, and to retain its grosser parts. For I have found some earths to yeeld by distillation a liquor very far from being inodorous or insipid; and 'tis a known observation that most kinds of fat earth kept covered from the rain, and hindred from spending themselves in the production of vegetables, will in time become impregnated with salt petre.

But I must remember that the water and earths I ought here to speak of, are such as are separated from mixt bodies by the fire; and therefore to restrain my discourse to such, I shall tell you, that we see the phlegme of vitriol (for instance) is a very effectual remedie against burnes; and I know a very famous and experienced physitian, whose unsuspected secret (himself confessed to me) it is, for the discussing of hard and obstinate tumours. The phlegme of vinegar, though drawn exceeding leasurely in a digesting furnace, I have purposely made tryal of; and sometimes found it able to draw, though slowly, a saccharine sweetness out of lead; and as I remember by long digestion, I dissolved corals in it. The phlegme of the sugar of saturne is said to have very peculiar properties. Divers eminent chymists teach, that it will dissolve pearls, which being precipitated by the spirit of the same concrete are thereby (as they say) rendred volatile; which has been confirmed to me, upon his own observation, by a person of great veracity. The phlegme of wine, and indeed divers other liquors that are indiscriminately condemned to be cast away as phlegm, are endowed with qualities that make them differ both from mere water, and from each other; and whereas the chymists are pleased to call the *caput mortuum* of what they have distilled (after they have by affusion of water drawn away its salt) *terra damnata,* or earth, it may be doubted whether or no those earths are all of them perfectly alike: and it is scarce to be doubted, but that there are some of them which remain yet unreduced to an elementary nature. The ashes of wood deprived of all the salt, and bone-ashes, or calcined hartshorn, which refiners choose to make tests of, as freest from salt, seem unlike: and he that shall compare either of these insipid ashes to lime, and much more to the *calx* of talck, (though by the affusion of water they be exquisitely dulcifyed) will perhaps see cause to think them things of a somewhat differing nature. And it is evident in colcothar that the exactest calcination, followed by an exquisite dulcification, does not alwaies reduce the remaining body into elementary earth; for after the salt or vitriol (if the calcination have been too faint) is drawn out of the calcothar, the residue is not earth, but a mixt body, rich in medical vertues (as experience has informed me) and which Angelus Sala affirmes to be partly reducible into malleable copper; which I judge very probable; for though when I was making experiments upon colcothar, I was destitute of a furnace capable of giving a heat intense enough to bring such a calx to fusion; yet having conjectured that if colcothar abounded with that metal, *aqua fortis* would find it out there, I put some dulcified colcothar into that menstruum, and found the liquor according to my expectation presently coloured as highly as if it had been an ordinary solution of copper.

THE FIFTH PART

HERE Carneades making a pause, I must not deny (saies his friend to him) that I think you have sufficiently proved that these distinct substances which chymists are wont to obtain from mixt bodies, by their vulgar distillation, are not pure and simple enough to deserve, in rigor of speaking, the name of elements, or principles. But I suppose you have heard, that there are some modern spagyrists, who give out that they can by further and more skilfull purifications, so reduce the separated ingredients of mixt bodies to an elementary simplicity, that the oyles (for instance) extracted from all mixts shall as perfectly resemble one another, as the drops of water do.

If you remember (replies Carneades) that at the beginning of our conference with Philoponus, I declared to him before the rest of the company, that I would not engage myself at present to do any more than examine the usual proofs alledged by chymists, for the vulgar doctrine of their three hypostatical principles; you will easily perceive that I am not obliged to make answer to what you newly proposed; and that it rather grants, than disproves what I have been contending for: since by pretending to make so great a change in the reputed principles that distillation affords the common spagyrists, 'tis plainly enough presupposed, that before such artificial depurations be made, the substances to be made more simple were not yet simple enough to be looked upon as elementary; wherefore in case the artists you speak of could perform what they give out they can, yet I should not need to be ashamed of having questioned the vulgar opinion touching the *tria prima*. And as to the thing itself, I shall freely acknowledge to you, that I love not to be forward in determining things to be impossible, till I know and have considered the means by which they are proposed to be effected. And therefore I shall not peremptorily deny either the possibility of what these

154

artists promise, or my assent to any just inference; however destructive to my conjectures, that may be drawn from their performances. But give me leave to tell you withall, that because such promises are wont (as experience has more than once informed me) to be much more easily made, than made good by chymists, I must withhold my belief from their assertions, till their experiments exact it; and must not be so easie as to expect beforehand, an unlikely thing upon no stronger inducements than are yet given me: besides that I have not yet found by what I have heard of these artists, that though they pretend to bring the several substances into which the fire has divided the concrete, to an exquisite simplicity, they pretend also to be able by the fire to divide all concretes, minerals, and others, into the same number of distinct substances. And in the meantime I must think it improbable, that they can either truly separate as many differing bodies from gold (for instance) or ostiocolla, as we can do from wine, or vitriol; or that the mercury (for example) of gold or saturn would be perfectly of the same nature with that of hartshorn; and that the sulphur of antimony would be but numerically different from the distilled butter or oyle of roses.

But suppose (saies Eleutherius) that you should meet with chymists, who would allow you to take in earth and water into the number of the principles of mixt bodies; and being also content to change the ambiguous name of mercury for that more intelligible one of spirit, should consequently make the principles of compound bodies to be five; would you not think it something hard to reject so plausible an opinion, only because the five substances into which the fire divides mixt bodies are not exactly pure, and homogeneous? For my part (continues Eleutherius) I cannot but think it somewhat strange, in case this opinion be not true, that it should fall out so luckily, that so great a variety of bodies should be analyzed by the fire into just five distinct substances; which so little differing from the bodies that bear those names, may so plausibly be called oyle, spirit, salt, water, and earth.

The opinion you now propose (answers Carneades) being another than that I was engaged to examine, it is not requisite for me to debate at this present; nor should I have leasure to do it thoroughly. Wherefore I shall only tell you in general, that though I think this opinion in some respects more defensible than that of the vulgar chymists; yet you may easily enough learn from the past discourse what may be thought of it: since many of the objections made against the vulgar doctrine of the chymists seem, without much alteration, employable against this hypothesis also. For, besides that this doctrine does as well as the other take it for granted, (what is not easie to be proved) that the fire is the true and adequate analyzer of bodies, and that all the distinct substances obtainable from a mixt body by the fire, were so pre-existent in it, that they were but extricated from each other by the analysis; besides that this opinion, too, ascribes to the productions of the fire an elementary simplicity, which I have shewn not to belong to them; and besides that this doctrine is lyable to some of the other difficulties, wherewith that of the *tria prima* is incumbered; besides all this, I say, this quinary number of elements, (if you pardon the expression) ought at least to have been restrained to the generality of animal and vegetable bodies, since not only among these there are some bodies, (as I formerly argued) which, for ought yet has been made to appear, do consist, either of fewer or more similar substances than precisely five. But in the mineral kingdom, there is scarce one concrete that has been evinced to be adequately divisible into such five principles or elements, and neither more nor lesse, as this opinion would have every mixt body to consist of.

And this very thing (continues Carneades) may serve to take away or lessen your wonder, that just so many bodies as five should be found upon the resolution of concretes. For since we find not that the fire can make any such analysis (into five elements) of metals and other mineral bodies whose texture is more strong and permanent, it remains that the five substances under consideration be obtained from vegetable and animal bodies, which (probably by reason of their looser contexture) are capable of being distilled. And as to such bodies, 'tis natural enough, that, whether we suppose that there are, or are not, precisely five elements, there should ordinarily occur in the dissipated parts a five-fold diversity of scheme (if I may so speak): for if the parts do not remain all fixed, as in gold, calcined talck, etc. nor all ascend, as in the sublimation of brimstone, camphire, etc. but after their dissipation do associate themselves into new schemes of matter; it is very likely, that they will by the fire be divided into fixed and volatile (I mean, in reference to that degree of heat by which they are distilled) and those volatile parts will, for the most part, ascend either in a dry forme, which chymists are pleased to call, if they be tasteless, flowers; if sapid, volatile salt; or in a liquid forme. And this liquor must be either inflamable, and so pass for oyl, or not inflamable, and yet subtile and pungent, which may be called spirit; or else strengthless or insipid, which may be named phlegme, or water. And as for the fixt part, or *caput mortuum*, it will most commonly consist of corpuscles, partly soluble in water, or sapid, (especially if the saline parts were not so volatile, as to fly away before) which make up its fixt salt; and partly insoluble and insipid, which therefore seems to challenge the name of earth. But although upon this ground one might easily enough have foretold, that the differing substances obtained from a perfectly mixt body by the fire would for the most part be reducible to the five newly mentioned states of matter; yet it will not presently follow, that these five distinct substances were simple and primogeneal bodies, so pre-existent in the concrete that the fire does but take them asunder. Besides that it does not appear, that all mixt bodies (witness, gold, silver, mercury, etc.) nay nor perhaps all vegetables, which may appear by what we said above of *camphire*, *benzoin*, etc., are resoluble by fire into just such differing schemes of matter. Nor will the experiments formerly alledged permit us to look upon these separated substances as elementary, or uncompounded. Neither will it be a sufficient argument of their being bodies that deserve the names which chymists are pleased to give them, that they have an analogy in point of consistence, or either volatility or fixtness, or else some other obvious quality, with the supposed principles, whose names are ascribed to them. For, as I told you above, notwithstanding this resemblance in some one quality, there may be such a disparity in others, as may be more fit to give them differing appellations, than the resemblance is to give them one and the same. And indeed it seems but somewhat a gross way of judging of the nature of bodies, to conclude without scruple, that those must be of the same nature that agree in some such general quality, as fluidity, dryness, volatility, and the like: since each of those qualities, or states of matter, may comprehend a great variety of bodies, otherwise of a very differing nature; as we may see in the calxes of gold, of vitriol, and of Venetian talck, compared with common ashes, which yet are very dry, and fixed by the vehemence of the fire, as well as they. And as we may likewise gather from what I have formerly observed, touching the spirit of boxwood, which though a volatile, sapid, and not inflamable liquor, as well as the spirits of hartshorn, of blood and others, (and therefore has been hitherto called, the spirit, and esteemed for one of the principles of the wood that affords it) may yet, as I told you, be subdivided into two liquors, differing from one another, and one of them at least, from the generality of other chymical spirits.

But you may yourself, if you please, (pursues Carneades) accomodate to the hypothesis you proposed what other particulars you shall think applicable to it, in the foregoing discourse. For I think it unseasonable for me to medle now any further with a controversie, which since it does not now belong to me, leaves me at liberty to take my own time to declare myself about it.

Eleutherius perceiving that Carneades was somewhat unwilling to spend any more time upon the debate of this opinion, and having perhaps some thoughts of taking hence a rise to make him discourse it more fully another time, thought not fit as then to make any further mention to him of the proposed opinion, but told him;

I presume I need not mind you, Carneades, that both the patrons of the ternary number of principles, and those that would have five elements, endeavour to back their experiments with a specious reason or two; and especially some of those embracers of the opinion last named (whom I have conversed with, and found them learned men) assigne this reason of the necessity of five distinct elements; that otherwise mixt bodies could not be so compounded and tempered as to obtain a due consistence and competent duration. For salt (say they) is the basis of solidity; and permanency in compound bodies, without which the other four elements might indeed be variously and loosly blended together, but would remain incompacted; but that salt might be dissolved into minute parts, and conveyed to the other substances to be compacted by it, and with it, there is a necessity of water. And that the mixture may not be too hard and brittle, a sulphureous or oyly principle must intervene to make the mass more tenacious; to this a mercurial spirit must be superadded; which by its activity may for a while permeate, and as it were leaven the whole mass, and thereby promote the more exquisite mixture and incorporation of the ingredients. To all which (lastly) a portion of earth must be added, which by its dryness and porosity may soak up part of that water wherein the salt was dissolved, and eminently concurr with the other ingredients to give the whole body the requisite consistence.

I perceive (saies Carneades smiling) that if it be true, as 'twas lately noted from the proverb, " That good wits have bad memories," you have that title, as well as a better, to a place among the good wits. For you have already more than once forgot, that I declared to you that I would at this conference examine only the experiments of my adversaries, not their speculative reasons. Yet 'tis not (subjoynes Carneades) for fear of medling with the argument you have proposed, that I decline the examining it at present. For if when we are more at leasure, you shall have a mind that we may solemnly consider of it together; I am confident we shall scarce finde it insoluble.   And in the meantime we may observe, that such a way of arguing may, it seems, be speciously accommodated to differing hypotheses.   For I find that Beguinus, and other assertors of the *tria prima*, pretend to make out by such a way, the requisiteness of their salt, sulphur and mercury, to constitute mixt bodies, without taking notice of any necessity of an addition of water and earth.

And indeed neither sort of chymists seem to have duly considered how great variety there is in the textures and consistences of compound bodies; and how little the consistence and duration of many of them seem to accommodate and be explicable by the proposed notion. And not to mention those almost incorruptible substances obtainable by the fire, which I have proved to be some-what compounded, and which the chymists will readily grant not to be perfectly mixt bodies: (not to mention these, I say) if you will but recall to mind some of those experiments, whereby I shewed you that out of common water only mixt bodies (and even living ones) of very differing consistences, and resoluble by fire into as many principles as other bodies acknowledged to be perfectly mixt; may be produced if you do this, I say, you will not, I suppose, be averse from believing, yet nature by a convenient disposition of the minute parts of a portion of matter may contrive bodies durable enough, and of this, or that, or the other consistence, without being obliged to make use of all, much less of any determinate quantity of each of the five elements, or of the three principles to compound such bodies of.   And I have (pursues Carneades) something wondered, chymists should not consider, that there is scarce any body in nature so permanent and indissoluble as glass; which yet them-selves teach us may be made of bare ashes, brought to fusion by the mere violence of the fire; so that, since ashes are granted to consist but of pure salt and simple earth, sequestred from all the other principles or elements, they must acknowledge, that even art itself can of two elements only, or, if you please, one principle and one element, compound a body more durable than almost any in the world. Which being undeniable, how will they prove that nature cannot compound mixt bodies, and even durable ones, under all the five elements or material principles.

But to insist any longer on this occasional disquisition, touching their opinion that would establish five elements, were to remember as little as you did before, that the debate of this matter is no part of my first undertaking; and consequently, that I have already spent time enough in what I look back upon but as a digression, or at best an excursion.

And thus, Eleutherius, (saies Carneades) having at length gone through the four considerations I proposed to discourse unto you, I hold it not unfit, for fear my having insisted so long on each of them may have made you forget their series, briefly to repeat them by telling you, that

Since, in the first place, it may justly be doubted whether or no the fire be, as chymists suppose it, the genuine and universal resolver of mixt bodies;

Since we may doubt, in the next place, whether or no all the distinct substances that may be obtained from a mixt body by the fire were pre-existent there in the formes in which they were separated from it;

Since also, though we should grant the substances separable from mixt bodies by the fire to have been their component ingredients, yet the number of such substances does not appear the same in all mixt bodies; some of them being resoluble into more differing substances than three, and others not being resoluble into so many as three;

And since, lastly, those very substances that are thus separated are not for the most part pure and elementary bodies, but new kinds of mixts;

Since, I say, these things are so, I hope you will allow me to inferr, that the vulgar experiments (I might perchance have added, the arguments too) wont to be alledged by chymists to prove, that their three hypostatical principles do adequately compose all mixt bodies, are not so demonstrative as to induce a wary person to acquiesce in their doctrine, which, till they explain and prove it better, will by its perplexing darkness be more apt to puzzle than satisfy considering men, and will to them appear incumbered with no small difficulties.

And from what has been hitherto deduced (continues Carneades) we may learn, what to judge of the common practice of those chymists, who because they have found that diverse compound bodies (for it will not hold in all) can be resolved into, or rather can be brought to afford two or three differing substances more than the soot and ashes, whereinto the naked fire commonly divides them in our chymnies, cry up their own sect for the invention of a new philosophy, some of them, as Helmont, etc. styling themselves philosophers by the fire; and the most part not only ascribing, but as far as in them lies, engrossing to those of their sect the title of PHILOSOPHERS.

But alas, how narrow is this philosophy, that reaches but to some of those compound bodies, which we find but upon, or in the crust or outside of our terrestrial globe, which is itself but a point in comparison of the vast extended universe, of whose other and greater parts the doctrine of the *tria prima* does not give us an account! For what does it teach us, either of the nature of the sun, which astronomers affirme to be eight-score and odd times bigger than the whole earth? or of that of those numerous fixt starrs, which, for ought we know, would very few, if any of them, appear inferiour in bulke and brightness to the sun, if they were as near us as he? What does the knowing that salt, sulphur and mercury, are the principles of mixt bodies, informe us of the nature of that vast, fluid, and ætherial substance, that seems to make up the interstellar, and consequently much the greatest part of the world? for as for the opinion commonly ascribed to Paracelsus, as if he would have not only the four peripatetick elements, but even the celestial parts of the universe to consist of his three principles, since the modern chymists themselves have not thought so groundless a conceit worth their owning, I shall not think it worth my confuting.

But I should perchance forgive the hypothesis I have been all this while examining, if, though it reaches but to a very little part of the world, it did at least give us a satisfactory account of those things to which 'tis said to reach. But find not, that it gives us any other than a very imperfect information even about mixt bodies themselves: for how will the knowledge of the *tria prima* discover to us the reason, why the loadstone drawes a needle, and disposes it to respect the poles, and yet seldom precisely points at them? How will this hypothesis teach us how a chick is formed in the egge, or how the seminal principles of mint, pompions, and other vegetables, that I mentioned to you above, can fashion water into various plants, each of them endowed with its peculiar and determinate shape, and with divers specifick and discriminating qualities? How does this hypothesis shew us, how much salt, how much sulphur, and how much mercury must be taken to make a chick or a pompion? and if we know that: what principle is it that manages these ingredients, and contrives (for instance) such liquors as the white and yolk of an egge into such a variety of textures as is requisite to fashion the bones, veines, arteries, nerves, tendons, feathers, blood, and other parts of a chick; and not only to fashion each limbe, but to connect them altogether, after that manner that is most congruous to the perfection of the animal which is to consist of them? For to say, that some more fine and subtile part of either or all the hypostatical principles is the director in all this business, and the architect of all this elaborate structure, is to give one occasion to demand again, what proportion and way of mixture of the *tria prima* afforded this architectonick spirit, and what agent made so skilful and happy a mixture? And the answer to this question, if the chymists will keep themselves within their three principles, will be lyable to the same inconvenience, that the answer to the former was. And if it were not to intrench upon the theame of a friend of ours here present, I could easily prosecute the imperfections of the vulgar chymists philosophy, and shew you, that by going about to explicate by their three principles, I say not, all the abstruse properties of mixt bodies, but even such obvious and more familiar phænomena as *fluidity* and *firmness*, the colours and figures of stones, minerals, and other compound bodies, the nutrition of either plants or animals, the gravity of gold or quicksilver compared with wine or spirit of wine; by attempting, I say, to render a reason of these (to omit a thousand others as difficult to account for) from any proportion of the three simple ingredients, chymists will be much more likely to discredit themselves and their hypothesis, than satisfy an intelligent inquirer after truth.

But (interposes Eleutherius) this objection seems no more than may be made against the four peripatetick elements. And indeed almost against any other hypothesis, that pretends by any determinate number of material ingredients to render a reason of the phænomena of nature. And as for the use of the chymical doctrine of the three principles, I suppose you need not be told by me, that the great champion of it, the learned Sennertus, assignes this noble use of the *tria prima*, that from them, as the nearest and most proper principles, may be deduced and demonstrated the properties which are in mixt bodies, and which cannot be proximately (as they speak) deduced from the elements. And this, saies he, is chiefly apparent, when we inquire into the properties and faculties of medicines. And I know (continues Eleutherius) that the person you have assumed, of an opponent of the hermetick doctrine, will not so far prevaile against your native and wonted equity, as to keep you from acknowledging that philosophy is much beholden to the notions and discoveries of chymists.

If the chymists you speak of (replyes Carneades) had been so modest, or so discreet, as to propose their opinion of the *tria prima*, but as a notion useful among others, to increase humane knowledge, they had deserved more of our thanks, and less of our opposition; but since the thing, that they pretend, is not so much to contribute a notion toward the improvement of philosophy, as to make this notion (attended by a few less considerable ones) pass for a new philosophy itself; nay, since they boast so much of this phancie of theirs, that the famous Quercetanus scruples not to write, that if his most certain doctrine of the three principles were sufficiently learned, examined, and cultivated, it would easily dispel all the darkness that benights our minds, and bring in a clear light, that would remove all difficulties: this school affording theorems and axioms irrefragable, and to be admitted without dispute by impartial judges; and so useful withal, as to exempt us from the necessity of having recourse, for want of the knowledge of causes, to that sanctuary of the ignorant, occult qualities; since I say, this domestick notion of the chymists is so much over-valued by them, I cannot think it unfit, they should be made sensible of their mistake; and be admonished to take in more fruitful and comprehensive principles, if they mean to give us an account of the phænomena of nature; and not confine themselves, and (as far as they can) others, to such narrow principles, as I fear will scarce enable them to give an account (I mean an intelligible one) of the tenth part (I say not) of all the phænomena of nature; but even of all such as by the Leucippian or some of the other sorts of principles may be plausibly enough expli-cated. And though I be not unwilling to grant, that the incompetency I impute to the chymical hypothesis is but the same which may be objected against that of the four elements, and divers other doctrines that have been maintained by learned men; yet since 'tis the chymical hypothesis only which I am now examining, I see not why, if what I impute to it be a real inconvenience, either it should cease to be so, or I should scruple to object it, because other theories are lyable thereunto, as well as the hermetical. For I know not why a truth should be thought less a truth for the being fit to overthrow variety of errors.

I am obliged to you (continues Carneades, a little smiling) for the favourable opinion you are pleased to express of my equity, if there be no designe in it. But I need not be tempted by an artifice, or invited by a complement, to acknowledge the great service that the labours of chymists have done the lovers of useful learning; nor even on this occasion shall their arrogance hinder my gratitude. But since we are as well examining the truth of their doctrine, as the merit of their industry, I must in order to the investigation of the first, continue a reply, to talk at the rate of the part I have assumed; and tell you, that when I acknowledge the usefulness of the labours of spagyrists to natural philosophy, I do it upon the score of their experiments, not upon that of their speculations; for it seems to me, that their writings, as their furnaces, afford as well smoak as light; and do little less obscure some subjects, than they illustrate others. And though I am unwilling to deny, that 'tis difficult for a man to be an accomplisht naturalist, that is a stranger to chymistry; yet I look upon the common operations and practices of chymists, almost as I do on the letters of the alphabet, without whose knowledge 'tis very hard for a man to become a philosopher; and yet that knowledge is very far from being sufficient to make him one.

But (saies Carneades, resuming a more serious look) to consider a little more particularly what you alledge in favour of the chymical doctrine of the *tria prima*, though I shall readily acknowledge it not to be unuseful, and that the divisers and embracers of it have done the commonwealth of learning some service, by helping to destroy that excessive esteem, or rather veneration, wherewith the doctrine of the four elements was almost as generally, as undeservedly entertained; yet what has been alledged concerning the usefulness of the *tria prima*, seems to me liable to no contemptible difficulties.

And first, as for the very way of probation, which the more learned and more sober champions of the chymical cause employ to evince the chymical principles in mixt bodies, it seems to me to be farr enough from being convincing. This grand and leading argument, your Sennertus himself, who layes great weight upon it, and tells us, that the most learned philosophers employ this way of reasoning to prove the most important things, proposes thus: " Ubicunque (saies he) pluribus eædem affectiones et qualitates insunt, per commune quoddam principium insint necesse est, sicut omnia sunt gravia propter terram, calida propter ignem. At colores, odores, sapores, esse φλογιστὲν, et similia alia, mineralibus, metallis, gemmis, lapidibus, plantis, animalibus insunt. Ergo per commune aliquod principium, et subjectum, insunt. At tale principium non sunt elementa. Nullam enim habent ad tales qualitates producendas potentiam. Ergo alia principia, unde fluant, inquirenda sunt."

In the recital of this argument, (saies Carneades) I therefore thought fit to retain the language wherein the author proposes it, that I might also retaine the propriety of some Latine termes, to which I do not readily remember any that fully answer in English. But as for the argumentation itself, 'tis built upon a precarious supposition, that seems to me neither demonstrable nor true; for, how does it appear that where the same quality is to be met with in many bodies, it must belong to them upon the account of some one body whereof they all partake? (For that the major of our authors argument is to be understood of the material ingredients of bodies, appears by the instances of earth and fire he annexes to explain it.) For to begin with that very example which he is pleased to alledge for himself; how can he prove, that the gravity of all bodies proceeds from what they participate of the element of earth? Since we see, that not only common water, but the more pure distilled rain water is heavy; and quicksilver is much heavier than earth itself; though none of my adversaries has yet proved, that it contains any of that element. And I the rather make use of this example of quicksilver, because I see not how the assertors of the elements will give any better account of it than the chymists. For if it be demanded how it comes to be fluid, they will answer, that it participates much of the nature of water. And indeed, according to them, water may be the predominant element in it, since we see, that severall bodies, which by distillation afford liquors that weigh more than their *caput mortuum*, do not yet consist of liquor enough to be fluid. Yet if it be demanded how quicksilver comes to be so heavy, then 'tis replyed, that 'tis by reason of the earth that abounds in it; but since, according to them, it must consist also

of air, and partly of fire, which they affirme to be light
elements, how comes it that it should be so much heavier
than earth of the same bulk, though to fill up the porosities
and other cavities it be made up into a mass or paste
with water, which itself they allow to be a heavy element.
But to returne to our spagyrists, we see that chymical
oyles and fixt salts, though never so exquisitely purifyed
and freed from terrestrial parts, do yet remain ponderous
enough. And experience has informed me, that a pound
(for instance) of some of the heaviest woods, as guajacum,
that will sinke in water, being burnt to ashes will yeeld
a much less weight of them (whereof I found but a small
part to be alcalizate) than much lighter vegetables: as
also that the black charcoal of it will not sink as did the
wood, but swim; which argues that the differing gravity
of bodies proceeds chiefly from the particular texture,
as is manifest in gold, the closest and compactest of
bodies, which is many times heavier than we can possibly
make any parcel of earth of the same bulk. I will not
examine, what may be argued touching the gravity or
quality analogous thereunto, of even celestial bodies,
from the motion of the spots about the sun, and from the
appearing equality of the supposed seas in the moon;
nor consider how little those phænomena would agree
with what Sennertus presumes concerning gravity. But
further to invalidate his supposition, I shall demand, upon
what chymical principle fluidity depends? And yet
fluidity is, two or three perhaps excepted, the most diffused
quality of the universe, and far more general than almost
any other of those that are to be met with in any of the
chymical principles, or Aristotelian elements; since not
only the air, but that vast expansion we call heaven,
in comparison of which our terrestrial globe (supposing
it were all solid) is but a point; and perhaps too the sun
and the fixt stars are fluid bodies. I demand also, from
which of the chymical principles motion flowes; which
yet is an affection of matter much more general than any
that can be deduced from any of the three chymical
principles. I might ask the like question concerning
light, which is not only to be found in the kindled sulphur

of mixt bodies but (not to mention those sorts of rotten woods, and rotten fish that shine in the dark) in the tails of living glow-wormes, and in the vast bodies of the sun and stars. I would gladly also know, in which of the three principles the quality, we call sound, resides as in its proper subject; since either oyl falling upon oyle, or spirit upon spirit, or salt upon salt, in a great quantity, and from a considerable height, will make a noise, or if you please, create a sound, and (that the objection may reach the Aristotelians) so will also water upon water, and earth upon earth. And I could name other qualities to be met with in divers bodies, of which I suppose my adversaries will not in haste assign any subject, upon whose account in must needs be, that the quality belongs to all the other several bodies.

And, before I proceed any further, I must here invite you to compare the supposition we are examining, with some other of the chymical tenents. For, first they do in effect teach, that more than one quality may belong to, and be deduced from, one principle. For, they ascribe to salt, tastes, and the power of coagulation; to sulphur, as well odours as inflamableness; and some of them ascribe to mercury, colours; as all of them do effumability, as they speak. And on the other side, it is evident that volatility belongs in common to all the three principles, and to water too. For 'tis manifest that chymical oyles are volatile; that also divers salts, emerging upon the analysis of many concretes, are very volatile, is plain from the fugitiveness of salt, of hartshorn, flesh, etc. ascending in the distillation of those bodies. How easily water may be made to ascend in vapours, there is scarce anybody that has not observed. And as for what they call the mercurial principle of bodies, that is so apt to be raised in the form of steam, that Paracelsus and others define it by that aptness to fly up; so that (to draw that inference by the way) it seems not that chymists have been accurate in their doctrine of qualities, and their respective principles, since they both derive several qualities from the same principle, and must ascribe the same quality to almost all their principles and other

bodies besides. And thus much for the first thing taken
for granted, without sufficient proof, by your Sennertus:
and to add that upon the by (continues Carneades) we
may hence learn what to judge of the way of argumentation, which that fierce champion of the Aristotelians
against the chymists, Anthonius Guntherus Billichius
employes, where he pretends to prove against Beguinus,
that not only the four elements do immediately concurr
to constitute every mixt body, and are both present in it,
and obtainable from it upon its dissolution; but that in
the *tria prima* themselves, whereinto chymists are wont
to resolve mixt bodies, each of them clearly discovers
itself to consist of four elements. The ratiocination itself
(pursues Carneades) being somewhat unusual, I did the
other day transcribe it, and (saies he, pulling a paper
out of his pocket) it is this. " Ordiamur, cum Beguino,
à ligno viridi, quod si concrematur, videbis in sudore
aquam, in fumo aerem, inflamma et prunis ignem, terram
in cineribus: quod si Beguino placuerit ex eo colligere
humidum aquosum, cohibere humidum oleaginosum,
extrahere ex cineribus salem; ego ipsi in unoquoque
horum seorsim quatuor elementa ad oculum demonstrabo,
eodem artificio quo in ligno viridi ea demonstravi.
Humorem aquosum admoveho igni. Ipse aquam ebullire
videbit, in vapore aerem conspiciet, ignem sentiet in
æstu, plus minus terræ in sedimento apparebit. Humor
porro oleaginosus aquam humiditate et fluiditate per se,
accensus vero ignem flamma prodit, fumo aerem, fuligine,
nidore et amurca terram. Salem denique ipse Beguinus
siccum vocat et terrestrem, qui tamen nec fusus aquam,
nec caustica vi ignem celare potest; ignis vero violentia
in halitus versus nec ab aere se alienum esse demonstrat;
idem de lacte, de ovis, de semine lini, de garyophyllis,
de nitro, de sale marino, denique de antimonio, quod
fuit de ligno viridi judicium; eadem de illorum partibus,
quas Beguinus adducit, sententia, quæ de viridis ligni
humore aquoso, quæ de liquore ejusdem oleoso, quæ
de sale fuit."

    This bold discourse (resumes Carneades, putting up
again his paper) I think it were not very difficult to con-

fute, if his arguments were as considerable, as our time
will probably prove short for the remaining and more
necessary part of my discourse; wherefore referring you
for an answer to what was said concerning the dissipated
parts of a burnt piece of green wood, to what I told
Themistius on the like occasion, I might easily shew you,
how slightly and superficially our Guntherus talks of the
dividing the flame of green wood into his four elements;
when he makes that vapour to be air, which being caught
in glasses and condensed, presently discovers itself to have
been but an aggregate of innumerable very minute drops
of liquor; and when he would prove the phlegmes being
composed of fire by that heat which is adventitious to the
liquor, and ceases upon the absence of what produced it
(whether that be an agitation proceeding from the motion
of the external fire, or the presence of a multitude of
igneous atomes pervading the pores of the vessel, and
nimbly permeating the whole body of the water) I might,
I say, urge these and divers other weaknesses of his discourse. But I will rather take notice of what is more
pertinent to the occasion of this digression, namely, that
taking it for granted, that fluidity (with which he unwarily
seems to confound humidity) must proceed from the
element of water, he makes a chymical oyle to consist of
that elementary liquor; and yet in the very next words
proves, that it consists also of fire, by its inflamability;
not remembring that exquisitely pure spirit of wine is
both more fluid than water itself, and yet will flame all
away without leaving the least aqueous moisture behind
it; and without such an *amurca* and soot as he would
deduce the presence of earth from. So that the same
liquor may according to his doctrine be concluded by its
great fluidity to be almost all water; and by its burning
all away to be all disguised fire. And by the like way of
probation our author would shew that the fixt salt of
wood is compounded of the four elements. For (saies he)
being turned by the violence of the fire into steames, it
shews itself to be of kin to air; whereas I doubt whether
he ever saw a true fixt salt (which to become so, must
have already endured the violence of an incinerating fire)

brought by the fire alone to ascend in the forme of exhalations; but I do not doubt that if he did, and had caught
those exhalations in convenient vessels, he would have
found them as well as the steames of common salt, etc.
of a saline, and not an aëreal nature. And whereas our
author takes it also for granted, that the fusibility of salt
must be deduced from water it is indeed so much the
effect of heat variously agitating the minute parts of a
body, without regard to water, that gold (which by its
being the heaviest and fixtest of bodies, should be the
most earthy) will be brought to fusion by a strong fire;
which sure is more likely to drive away, than increase its
aqueous ingredient, if it have any; and on the other side,
for want of a sufficient agitation of its minute parts, ice
is not fluid, but solid; though he presumes also that the
mordicant quality of bodies must proceed from a fiery
ingredient; whereas, not to urge that the light and
inflamable parts, which are the most likely to belong
to the element of fire, must probably be driven away by
that time the violence of the fire has reduced the body
to ashes; not to urge this, I say, nor that oyle of vitriol
which quenches fire, burnes the tongue and flesh of those
that unwarily taste or apply it, as a caustick doth, it is
precarious to prove the presence of fire in fixt salts from
their caustick power, unless it were first shewn, that all
the qualities ascribed to salts must be deduced from those
of the elements; which, had I time, I could easily manifest
to be no easy task. And not to mention that our author
makes a body, as homogeneous, as any he can produce for
elementary, belong both to water and fire, though it be
neither fluid nor insipid, like water; nor light and volatile,
like fire; he seems to omit in this anatomy the element
of earth, save that he intimates, that the salt may pass for
that: but since a few lines before, he takes ashes for earth,
I see not how he will avoid an inconsistency either betwixt
the parts of his discourse, or betwixt some of them and his
doctrine. For since there is a manifest difference betwixt
the saline and the insipid parts of ashes, I see not how
substances, that disagree in such notable qualities, can
be both said to be portions of an element, whose nature

requires that it be homogeneous, especially in this case where an analysis by the fire is supposed to have separated it from the admixture of other elements, which are confessed by most Aristotelians to be generally found in common earth, and to render it impure. And sure if when we have considered for how little a disparities sake the peripateticks make these symbolizing bodies, aire and fire, to be two distinct elements, we shall also consider that the saline part of ashes is very strongly tasted, and easily soluble in water; whereas the other part of the same ashes is insipid and indissoluble in the same liquor: not to add, that the one substance is opacous, and the other somewhat diaphanous, nor that they differ in divers other particulars; if we consider those things, I say, we shall hardly think that both these substances are elementary earth; and as to what is sometimes objected, that their saline taste is only an effect of incineration and adustion, it has been elsewhere fully replyed to, when proposed by Themistius, and where it has been proved against him, that however insipid earth may perhaps by additaments be turned into salt, yet 'tis not like it should be so by the fire alone: for we see that when we refine gold and silver, the violentest fires we can employ on them give them not the least relish of saltness. And I think Philoponus has rightly observed, that the ashes of some concretes contain very little salt if any at all; for refiners suppose that bone-ashes are free from it, and therefore make use of them for tests and cuppels, which ought to be destitute of salt, lest the violence of the fire should bring them to vitrification; and having purposely and heedfully tasted a cuppel made of only bone-ashes and fair water, which I had caused to be exposed to a very violent fire, actuated by the blast of a large pair of double bellows, I could not perceive that the force of the fire had imparted to it the least saltness, or so much as made it less insipid.

But (saies Carneades) since neither you nor I love repetitions, I shall not now make any of what else was urged against Themistius, but rather invite you to take notice with me, that when our authour, though a learned man, and one that pretends skill enough in chymistry to reforme the whole art, comes to make good his confident undertaking, to give us an ocular demonstration of the immediate presence of the four elements in the resolution of green wood, he is fain to say things that agree very little with one another. For about the beginning of that passage of his lately recited to you, he makes the sweat, as he calls it, of the green wood to be water, the smoak aire, the shining matter fire, and the ashes earth; whereas a few lines after, he will in each of these, nay (as I just now noted) in one distinct part of the ashes, shew the four elements. So that either the former analysis must be incompetent to prove that number of elements, since by it the burnt concrete is not reduced into elementary bodies, but into such as are yet each of them compounded of the four elements; or else these qualities, from which he endeavours to deduce the presence of all the elements in the fixt salt, and each of the other separated substances, will be but a precarious way of probation: especially if you consider, that the extracted alcali of wood, being, for ought appears, at least as similar a body, as any that the peripateticks can shew us, if its differing qualities must argue the presence of distinct elements, it will scarce be possible for them by any way they know of employing the fire upon any body, to shew that any body is a portion of a true element: and this recals to my mind, that I am now but in an occasional excursion, which aiming only to shew, that the peripateticks as well as the chymists take in our present controversie something for granted, which they ought to prove, I shall returne to my exceptions, where I ended the first of them, and further tell you, that neither is that the only precarious thing that I take notice of in Sennertus his argumentation; for when he inferrs, that because the qualities he mentions, as colours, smels, and the like, belong not to the elements, they therefore must to the chymical principles; he takes that for granted, which will not in haste be proved; as I might here manifest, but that I may by and by have a fitter opportunity to take notice of it. And thus much at present may suffice to have discoursed against the supposition, that almost every quality must have some δεκτικὸν πρῶτον, as they speak, some native receptacle, wherein as in its proper subject of inhesion it peculiarly resides; and on whose account that quality belongs to the other bodies, wherein it is to be met with. Now this fundamental supposition being once destroyed, whatsoever is built upon it, must fall to ruine of itself.

But I consider further, that chymists are (for ought I have found) far from being able to explicate by any of the *tria prima*, those qualities which they pretend to belong primarily unto it, and in mixt bodies to deduce from it. 'Tis true indeed, that such qualities are not explicable by the four elements; but it will not therefore follow that they are so by the three hermetical principles; and this is it that seems to have deceived the chymists, and is indeed a very common mistake amongst most disputants, who argue as if there could be but two opinions concerning the difficulty about which they contend; and consequently they inferr, that if their adversaries opinion be erroneous, their's must needs be the truth; whereas many questions, and especially in matters physiological, may admit of so many differing hypotheses, that 'twill be very inconsiderate and fallacious to conclude (except where the opinions are precisely contradictory) the truth of one from the falsity of another. And in our particular case 'tis no way necessary, that the properties of mixt bodies must be explicable either by the hermetical, or the Aristotelian hypothesis; there being divers other and more plausible waies of explaining them, and especially that, which deduces qualities from the motion, figure, and contrivance of the small parts of bodies; as I think might be shewn, if the attempt were as seasonable, as I fear it would be tedious.

I will allow then, that the chymists do not causelesly accuse the doctrine of the four elements of incompetency to explain the properties of compound bodies. And for this rejection of a vulgar error, they ought not to be denyed what praise men may deserve for exploding a doctrine whose imperfections are so conspicuous, that men needed but not to shut their eyes, to discover them.

But I am mistaken, if our hermetical philosophers themselves need not, as well as the peripateticks, have recourse
to more fruitfull and comprehensive principles than the
tria prima, to make out the properties of the bodies they
converse with. Not to accumulate examples to this
purpose (because I hope for a fitter opportunity to prosecute this subject) let us at present only point at colour,
that you may guess by what they say of so obvious and
familiar a quality, how little instruction we are to expect
from the *tria prima* in those more abstruse ones, which
they with the Aristotelians stile occult. For about
colours, neither do they at all agree among themselves,
nor have I met with any one, of which of the three
perswasions soever, that does intelligibly explicate them.
The vulgar chymists are wont to ascribe colours to
mercury; Paracelsus in divers places attributes them
to salt; and Sennertus, having recited their differing
opinions, dissents from both; and referrs colours rather
unto sulphur. But how colours do, nay, how they may,
arise from either of these principles, I think you will
scarce say that any has yet intelligibly explicated. And
if Mr. Boyle will allow me to shew you the experiments
which he has collected about colours, you will, I doubt
not, confess that bodies exhibite colours, not upon the
account of the predominancy of this or that principle in
them, but upon that of their texture, and especially the
disposition of their superficial parts, whereby the light
rebounding thence to the eye is so modified, as by differing
impressions variously to affect the organs of sight. I
might here take notice of the pleasing variety of colours
exhibited by the triangular glass (as 'tis wont to be called)
and demand, what addition or decrement of either salt,
sulphur, or mercury, befalls the body of the glass by being
prismatically figured; and yet 'tis known, that without
that shape it would not afford those colours as it does.
But because it may be objected, that these are not real,
but apparent colours; that I may not lose time in examining the distinction, I will alledge against the chymists, a
couple of examples of real and permanent colours drawn
from metalline bodies; and represent, that without the

addition of any extraneous body, quicksilver may by the fire alone, and that in glasse vessels, be deprived of its silver-like colour, and be turned into a red body; and from this red body without addition likewise may be obtained a mercury bright and specular as it was before; so that I have here a lasting colour generated and destroyed (as I have seen) at pleasure, without adding or taking away either mercury, salt, or sulphur; and if you take a clean and slender piece of hardened steel, and apply to it the flame of a candle at some little distance short of the point, you shall not have held the steel long in the flame, but you shall perceive divers colours, as yellow, red and blew, to appear upon the surface of the metal, and as it were run along in chase of one another towards the point; so that the same body, and that in one and the same part, may not only have a new colour produced in it, but exhibite successively divers colours within a minute of an hour, or thereabouts; and any of these colours may by removing the steel from the fire, become permanent, and last many years, and this production and variety of colours cannot reasonably be supposed to proceed from the accession of any of the three principles, to which of them soever chymists will be pleased to ascribe colours; especially considering, that if you but suddenly refrigerate that iron, first made red hot, it will be hardened and colourless again; and not only by the flame of a candle, but by any other equivalent heat conveniently applied, the like colours will again be made to appear and succeed one another, as at the first. But I must not any further prosecute an occasional discourse, though that were not so difficult for me to do, as I fear it would be for the chymists to give a better account of the other qualities, by their principles, than they have done of colours. And your Sennertus himself (though an author I much value) would I fear have been exceedingly puzled to resolve, by the *tria prima*, halfe that catalogue of problems, which he challenges the vulgar peripateticks to explicate by their four elements. And supposing it were true, that salt or sulphur were the principle to which this or that quality may be peculiarly referred, yet though he that teaches us this, teaches us something concerning that quality, yet he teaches us but something. For indeed he does not teach us that which can in any tolerable measure satisfie an inquisitive searcher after truth. For what is it to me to know, that such a quality resides in such a principle or element, whilst I remain altogether ignorant of the cause of that quality, and the manner of its production and operation? How little do I know more than any ordinary man of gravity, if I know but that the heaviness of mixt bodies proceeds from that of the earth they are composed of, if I know not the reason why the earth is heavy? And how little does the chymist teach the philosopher of the nature of purgation, if he only tells him that the purgative vertue of medicines resides in their salt: for, besides that this must not be conceded without limitation, since the purging parts of many vegetables extracted by the water wherein they are infused, are at most but such compounded salts (I mean mingled with oyle, and spirit, and earth, as tartar and divers other subjects of the vegetable kingdom afford) and since too that quicksilver precipitated either with gold, or without addition, into a powder, is wont to be strongly enough cathartical, though the chymists have not yet proved, that either gold or mercury have any salt at all, much less any that is purgative; besides this, I say, how little is it to me, to know that 'tis the salt of the rhubarb (for instance) that purges, if I find that it does not purge as salt; since scarce any elementary salt is in small quantity cathartical. And if I know not how purgation in general is effected in a humane body? In a word, as 'tis one thing to know a man's lodging, and another, to be acquainted with him; so it may be one thing to know the subject wherein a quality principally resides, and another thing to have a right notion and knowledge of the quality itself. Now that which I take to be the reason of this chymical deficiency, is the same upon whose account I think the Aristotelian and divers other theories incompetent to explicate the origine of qualities. For I am apt to think, that men will never be able to explain the phænomena of nature, while they endeavour to deduce them only from the presence and proportion of such or such material ingredients, and consider such ingredients or elements as bodies in a state of rest; whereas indeed the greatest part of the affections of matter, and consequently of the phænomena of nature, seems to depend upon the motion and the contrivance of the small parts of bodies. For 'tis by motion that one part of matter acts upon another; and 'tis, for the most part, the texture of the body upon which the moving parts strike, that modifies the motion or impression, and concurrs with it to the production of those effects which make up the chief part of the naturalists theme.

But (saies Eleutherius) methinks for all this, you have left some part of what I alledged in behalf of the three principles, unanswered. For all that you have said will not keep this from being a useful discovery, that since in the salt of one concrete, in the sulphur of another, and the mercury of a third, the medicinal vertue of it resides; that principle ought to be separated from the rest, and there the desired faculty must be sought for.

I never denyed (replies Carneades) that the notion of the *tria prima* may be of some use, but (continues he laughing) by what you now alledge for it, it will but appear that it is useful to apothecaries rather than to philosophers: the being able to make things operative being sufficient to those, whereas the knowledge of causes is the thing looked after by these. And let me tell you, Eleutherius, even this itself will need to be entertained with some caution.

For first, it will not presently follow, that if the purgative or other vertue of a simple may be easily extracted by water or spirit of wine, it resides in the salt or sulphur of the concrete; since unless the body hath before been resolved by the fire, or some other powerful agent, it will, for the most part, afford in the liquors I have named, rather the finer compounded parts of itself, than the elementary ones. As I noted before, that water will dissolve not only pure salts, but chrystals of tartar, gumme arabick, myrrhe and other compound bodies. As also spirit of wine will dissolve not only the pure sulphur of concretes, but likewise the whole substance of divers resinous bodies, as benzoin, the gummous parts of jalap, gumme lacca, and other bodies that are counted perfectly mixt. And we see that the extracts made either with water or spirit of wine are not of a simple and elementary nature, but masses consisting of the looser corpuscles, and finer parts of the concretes whence they are drawn; since by distillation they may be divided into more elementary substances.

Next, we may consider that even when there intervenes a chymical resolution by the fire, 'tis seldom in the saline or sulphureous principle, as such, that the desired faculty of the concrete resides; but, as that titular salt or sulphur is yet a mixt body, though the saline or sulphureous nature be predominant in it. For, if in chymical resolutions the separated substances were pure and simple bodies, and of a perfect elementary nature; no one would be indued with more specifick vertues, than another; and their qualities would differ as little as do those of water. And let me add this upon the by, that even eminent chymists have suffered themselves to be reprehended by me for their over great diligence in purifying some of the things they obtain by fire from mixt bodies. For though such compleatly purifyed ingredients of bodies might perhaps be more satisfactory to our understanding; yet others are often more useful to our lives; the efficacy of such chymical productions depending most upon what they retain of the bodies whence they are separated, or gain by the new associations of the dissipated among themselves; whereas if they were merely elementary, their uses would be comparatively very small; and the vertues of sulphurs, salts, or other such substances of one denomination, would be the very same.

And by the way (Eleutherius) I am inclined upon this ground to think, that the artificial resolution of compound bodies by fire does not so much enrich mankind, as it divides them into their supposed principles; as upon the score of its making new compounds by new combinations of the dissipated parts of the resolved body. For by this means the number of mixt bodies is considerably increased; and many of those new productions are endowed with useful qualities; divers of which they owe not to the body from which they were obtained, but to their newly acquired texture.

But thirdly, that which is principally to be noted is this, that as there are divers concretes, whose faculties reside in some one or other of those differing substances, that chymists call their sulphurs, salts, and mercuries, and consequently may be best obtained, by analyzing the concrete whereby the desired principles may be had severed or freed from the rest; so there are others wherein the noblest properties lodge not in the salt, or sulphur, or mercury, but depend immediately upon the form, or (if you will) result from the determinate structure of the whole concrete; and consequently they that go about to extract the vertues of such bodies, by exposing them to the violence of the fire, do exceedingly mistake, and take the way to destroy what they would obtain.

I remember that Helmont himself somewhere confesses, that as the fire betters some things and improves their vertues, so it spoyles others and makes them degenerate. And elsewhere he judiciously affirmes, that there may be sometimes greater vertue in a simple, such as nature has made it, than in anything that can by the fire be separated from it. And lest you should doubt whether he means by the vertues of things those that are medical; he has in one place this ingenuous confession; " Credo (saies he) simplicia in sua simplicitate esse sufficientia pro sanatione omnium morborum." Nay, Barthius, even in a comment upon Beguinus, scruples not to make this acknowledgment; " Valde absurdum est (saies he) ex omnibus rebus extracta facere, salia, quintas essentias; præsertim ex substantiis per se plane vel subtilibus vel homogeneis, quales sunt uniones, corallia, moscus, ambra, etc." Consonantly whereunto he also tells us, (and vouches the famous Platerus, for having candidly given the same advertisement to his auditors), that some things have greater vertues, and better suited to our humane nature, when unprepared, than when they have past the chymists fire; as we see, saies my author, in pepper; of which some grains swallowed perform more towards the relief of a distempered stomack, than a great quantity of the oyle of the same spice.

It has been (pursues Carneades) by our friend here present observed concerning salt-petre, that none of the substances into which the fire is wont to divide it, retaines either the taste, the cooling vertue, or some other of the properties of the concrete; and that each of those substances acquires new qualities not to be found in salt-petre itself. The shining property of the tayls of glow-worms does survive but so short a time the little animal made conspicuous by it, that inquisitive men have not scrupled publickly to deride Baptista Porta and others; who, deluded perhaps with some chymical surmises, have ventured to prescribe the distillation of a water from the tayles of glowwormes, as a sure way to obtain a liquor shining in the dark. To which I shall now add no other example than that afforded us by amber; which, whilst it remains an intire body, is endowed with an electrical faculty of drawing to itself feathers, strawes, and such like bodies; which I never could observe either in its salt, its spirit, its oyle, or in the body I remember I once made by the reunion of its divided elements; none of these having such a texture as the intire concrete. And however chymists boldly deduce such and such properties from this or that proportion of their component principles; yet in concretes that abound with this or that ingredient, 'tis not alwaies so much by vertue of its presence, nor its plenty, that the concrete is qualifyed to perform such and such effects; as upon the account of the particular texture of that and the other ingredients, associated after a determinate manner into one concrete: though possibly such a proportion of that ingredient may be more convenient than another for the constituting of such a body. Thus in a clock the hand is moved upon the dyal, the bell is struck, and the other actions belonging to the engine are performed, not because the wheeles are of brass or iron, or part of one metal and part of another, or because the weights are of lead, but by vertue of the size, shape, bigness, and co-aptation of the several parts; which would performe the same things though the wheels were of silver, or lead, or wood, and the weights of stone or clay; provided the fabrick or contrivance of the engine were the same: though it be not to be denied, that brass and steel are more convenient materials to make clock-wheels of than lead, or wood. And to let you see, Eleutherius, that 'tis sometimes at least, upon the texture of the small parts of a body, and not alwaies upon the presence, or recess, or increase, or decrement of any one of its principles, that it may loose some such qualities, and acquire some such others as are thought very strongly inherent to the bodies they reside in; I will add to what may from my past discourse be referred to this purpose, this notable example, from my own experience; That lead may without any additament, and only by various applications of the fire, lose its colour; and acquire sometimes a gray, sometimes a yellowish, sometimes a red, sometimes an amethystine colour; and after having past through these, and perhaps divers others, again recover its leaden colour, and be made a bright body. That also this lead, which is so flexible a metal, may be made as brittle as glasse, and presently be brought to be again flexible and malleable as before. And besides, that the same lead, which I find by microscopes to be one of the most opacous bodies in the world, may be reduced to a fine transparent glass; whence yet it may return to an opacous nature again; and all this, as I said, without the addition of any extraneous body, and merely by the manner and method of exposing it to the fire.

But (saies Carneades) after having already put you to so prolix a trouble, it is time for me to relieve you with a promise of putting speedily a period to it; and to make good that promise, I shall from all that I have hitherto discoursed with you, deduce but this one proposition by way of corollary. [*That it may as yet be doubted, whether or no there be any determinate number of elements ; or, if you please, whether or no all compound bodies, do consist of the same number of elementary ingredients or material principles.*]

This being but an inference from the foregoing discourse,

it will not be requisite to insist at large on the proofs of it; but only to point at the chief of them, and referr you for particulars to what has been already delivered.

In the first place, then, from what has been so largely discoursed, it may appear, that the experiments wont to be brought, whether by the common peripateticks, or by the vulgar chymists, to demonstrate, that all mixt bodies are made up precisely either of the four elements, or the three hypostatical principles, do not evince what they are alledged to prove. And as for the other common arguments, pretended to be drawn from reason in favour of the Aristotelian hypothesis (for the chymists are wont to rely almost altogether upon experiments) they are commonly grounded upon such unreasonable or precarious suppositions, that 'tis altogether as easie and as just for any man to reject them, as for those that take them for granted to assert them, being indeed all of them as indemonstrable as the conclusion to be inferred from them; and some of them so manifestly weak and proof-lesse; that he must be a very courteous adversary, that can be willing to grant them; and as unskilful a one, that can be compelled to do so.

In the next place, it may be considered, if what those patriarchs of the spagyrists, Paracelsus and Helmont, do on divers occasions positively deliver, be true; namely that the alkahest does resolve all mixt bodies into other principles than the fire, it must be decided which of the two resolutions (that made by the alkahest, or that made by the fire) shall determine the number of the elements, before we can be certain how many there are.

And in the meantime, we may take notice in the last place, that as the distinct substances whereinto the alkahest divides bodies, are affirmed to be differing in nature from those whereunto they are wont to be reduced by fire, and to be obtained from some bodies more in number than from some others; since he tells us, he could totally reduce all sorts of stones into salt only, whereas of a coal he had two distinct liquors. So although we should acquiesce in that resolution which is made by fire, we find not that all mixt bodies are thereby divided into

the same number of elements and principles; some concretes affording more of them than others do; nay and
sometimes this or that body affording a greater number
of differing substances by one way of management, than
the same yeelds by another.   And they that out of gold,
or mercury, or muscovy-glass, will draw me as many
distinct substances, as I can separate from vitriol, or
from the juice of grapes variously ordered, may teach me
that which I shall very thankfully learn.   Nor does it
appear more congruous to that variety that so much
conduceth to the perfection of the universe, that all
elemented bodies be compounded of the same number of
elements, than it would be for a language, that all its
words should consist of the same number of letters.

THE SIXTH PART

HERE Carneades having dispacht what he thought requisite to oppose against what the chymists are wont to alledge for proof of their three principles, paused a while, and looked about him, to discover whether it were time for him and his friend to rejoyne the rest of the company. But Eleutherius perceiving nothing yet to forbid them to prosecute their discourse a little further, said to his friend, (who had likewise taken notice of the same thing) I halfe expected, Carneades, that after you had so freely declared your doubting, whether there be any determinate number of elements, you would have proceeded to question whether there be any elements at all. And I confess it will be a trouble to me if you defeat me of my expectation; especially since you see the leasure we have allowed us may probably suffice to examine that paradox; because you have so largely deduced already many things pertinent to it, that you need but intimate how you would have them applyed, and what you would inferr from them.

Carneades having in vain represented that their leasure could be but very short, that he had already prated very long, that he was unprepared to maintain so great and so invidious a paradox, was at length prevailed with to tell his friend; Since, Eleutherius, you will have me discourse *ex tempore* of the paradox you mention, I am content, (though more perhaps to express my obedience, than my opinion) to tell you that (supposing the truth of Helmont's and Paracelsus's alkahestical experiments, if I may so call them) though it may seem extravagant, yet it is not absurd to doubt, whether, for ought has been proved, there be a necessity to admit any elements, or hypostatical principles, at all.

And, as formerly, so now, to avoid the needless trouble

of disputing severally with the Aristotelians and the chymists, I will address myself to oppose them I have last named, because their doctrine about the elements is more applauded by the moderns, as pretending highly to be grounded upon experience. And, to deal not only fairly but favourably with them, I will allow them to take in earth and water to their other principles. Which I consent to the rather, that my discourse may the better reach the tenents of the peripateticks; who cannot plead for any so probably as for those two elements; that of fire above the air being generally by judicious men exploded as an imaginary thing; and the air not concurring to compose mixt bodies as one of their elements, but only lodging in their pores, or rather replenishing, by reason of its weight and fluidity, all those cavities of bodies here below, whether compounded or not, that are big enough to admit it, and are not filled up with any grosser substance.

And, to prevent mistakes, I must advertize you, that I now mean by elements, as those chymists that speak plainest do by their principles, certain primitive and simple, or perfectly unmingled bodies; which not being made of any other bodies, or of one another, are the ingredients of which all those called perfectly mixt bodies are immediately compounded, and into which they are ultimately resolved: now whether there be any one such body to be constantly met with in all, and each, of those that are said to be elemented bodies, is the thing I now question.

By this state of the controversie you will, I suppose, guess, that I need not be so absurd, as to deny that there are such bodies as earth and water, and quicksilver, and sulphur: but I look upon earth and water, as component parts of the universe, or rather of the terrestrial globe, not of all mixt bodies. And though I will not peremptorily deny that there may sometimes either a running mercury, or a combustible substance be obtained from a mineral, or even a metal; yet I need not concede either of them to be an element in the sence above declared; as I shall have occasion to shew you by and by.

To give you then a brief account of the grounds I intend to proceed upon, I must tell you, that in matters of philosophy, this seems to me a sufficient reason to doubt of a known and important proposition, that the truth of it is not yet by any competent proof made to appear. And congruously hereunto, if I shew that the grounds, upon which men are perswaded that there are elements, are unable to satisfie a considering man, I suppose my doubts will appear rational.

Now the considerations that induce men to think, that there are elements, may be conveniently enough referred to two heads. Namely, the one, that it is necessary that nature make use of elements to constitute the bodies that are reputed mixt. And the other, that the resolution of such bodies manifests that nature had compounded them of elementary ones.

In reference to the former of these considerations, there are two or three things that I have to represent.

And I will begin with reminding you of the experiments I not long since related to you concerning the growth of pompions, mint, and other vegetables out of fair water. For by those experiments it seems evident, that water may be transmuted into all the other elements; from whence it may be inferred, both, that 'tis not everything chymists will call salt, sulphur, or spirit, that needs alwaies be a primordiate and ingenerable body. And, that nature may contex a plant (though that be a perfectly mixt concrete) without having all the elements previously presented to her to compound it of. And, if you will allow the relation I mentioned out of Mounsieur De Rochas to be true; then may not only plants, but animals and minerals too, be produced out of water. And however there is little doubt to be made, but that the plants my tryals afforded me, as they were like in so many other respects to the rest of the plants of the same denomination; so they would, in case I had reduced them to putrefaction, have likewise produced wormes or other insects, as well as the resembling vegetables are wont to do; so that water may, by various seminal principles, be successively transmuted into both plants and animals. And if we consider that not only men, but even sucking children are, but too often, tormented with solid stones; and that divers sorts of beasts themselves, (whatever Helmont against experience think to the contrary) may be troubled with great and heavy stones in their kidneys and bladders, though they feed but upon grass and other vegetables, that are perhaps but disguised water, it will not seem improbable that even some concretes of a mineral nature, may likewise be formed of water.

We may further take notice, that as a plant may be nourisht, and consequently may consist of common water; so may both plants and animals, (perhaps even from their seminal rudiments) consist of compound bodies, without having anything merely elementary brought them by nature to be compounded by them: this is evident in divers men, who whilst they were infants were fed only with milk, afterwards live altogether upon flesh, fish, wine, and other perfectly mixt bodies. It may be seen also in sheep, who on some of our English downs or plains, grow very fat by feeding upon the grass, without scarce drinking at all. And yet more manifestly in the magots that breed and grow up to their full bignesse within the pulps of apples, pears, or the like fruit. We see also, that dungs that abound with a mixt salt give a much more speedy increment to corn and other vegetables, than water alone would do: and it hath been assured me, by a man experienced in such matters, that sometimes when to bring up roots very early, the mould they were planted in was made over-rich, the very substance of the plant has tasted of the dung. And let us also consider a graft of one kind of fruit upon the upper bough of a tree of another kind. As (for instance) the scion of a pear upon a white-thorne; for there the ascending liquor is already altered, either by the root, or in its ascent by the bark, or both wayes, and becomes a new mixt body: as may appear by the differing qualities to be met with in the saps of several trees; as particularly, the medicinal vertue of the birch-water, which I have sometimes drunk upon Helmont's great and not undeserved commendation. Now the graft, being fastened to the stock, must necessarily nourish itself, and produce its fruit, only out of this compound juice prepared for it by the stock, being unable to come at any other aliment. And if we consider, how much of the vegetable he feeds upon may (as we noted above) remain in an animal; we may easily suppose, that the blood of that animal who feeds upon this, though it be a well constituted liquor, and have all the differing corpuscles, that make it up, kept in order by one presiding form, may be a strangly decompounded body, many of its parts being themselves decompounded. So little is it necessary that even in the mixtures which nature herself makes in animal and vegetable bodies, she should have pure elements at hand to make her compositions of.

Having said thus much touching the constitution of plants and animals, I might perhaps be able to say as much touching that of minerals, and even metals, if it were as easy for us to make experiment in order to the production of these, as of those. But the growth or increment of minerals being usually a work of excessively long time, and for the most part performed in the bowels of the earth, where we cannot see it, I must instead of experiments make use, on this occasion, of observations.

That stones were not all made at once, but that some of them are nowadayes generated, may (though it be denied by some) be fully proved by several examples, of which I shall now scarce alledge any other, than that famous place in France known by the name of Les Caves Goutieres, where the water falling from the upper parts of the cave to the ground does presently there condense into little stones, of such figures as the drops, falling either severally or upon one another, and coagulating presently into stone, chance to exhibit. Of these stones some ingenious friends of ours, that went a while since to visit that place, did me the favour to present me with some that they brought thence. And I remember that both that sober relator of his voyages, Van Linschoten, and another good author, inform us that in the diamond mines (as they call them) in the East-Indies, when having diged the earth, though to no great depth, they find diamonds and take them quite away; yet in a very few years they find in the same place new diamonds produced there since. From both which relations, especially the first, it seems probable that nature does not alwaies stay for divers elementary bodies, when she is to produce stones. And as for metals themselves, authors of good note assure us, that even they were not in the beginning produced at once altogether, but have been observed to grow; so that what was not a mineral or metal before, became one afterwards. Of this it were easie to alledge many testimonies of professed chymists. But that they may have the greater authority, I shall rather present you with a few borrowed from more unsuspected writers. " Sulphuris mineram (as the inquisitive P. Fallopius notes) quæ nutrix est caloris subterranei fabri seu archæi fontium et mineralium, infra terram citissimè renasci testantur historiæ metallicæ. Sunt enim loca è quibus si hoc anno sulphur effossum fuerit; intermissa fossione per quadriennium redeunt fossores et omnia sulphure, ut antea, rursus inveniunt plena." Pliny relates, " In Italiæ insula Ilva, gigni ferri metallum. Strabo multo expressius; effossum ibi metallum semper regenerari. Nam si effossio spatio centum annorum intermittebatur, et iterum illuc revertebantur, fossores reperisse maximam copiam ferri regeneratam." Which history not only is countenanced by Fallopius, from the income which the iron of that island yeelded the Duke of Florence in his time; but is mentioned more expressely to our purpose, by the learned Cesalpinus. " Vena (saies he) ferri copiosissima est in Italia; ob eam nobilitata Ilva Tyrrheni maris insula incredibili copia etiam nostris temporibus eam gignens: nam terra quæ eruitur, dum vena offoditur tota, procedente tempore in venam convertitur." Which last clause is therefore very notable, because from thence we may deduce, that earth, by a metalline plastick principle latent in it, may be in processe of time changed into a metal. And even Agricola himself, though the chymists complain of him as their adversary, acknowledges thus much and more; by telling us that at a town called Saga in Germany, they dig up iron in the fields, by sinking ditches two foot deep; and adding, that within

the space of ten years the ditches are digged again for
iron since produced, as the same metal is wont to be
obtained in Ilva. Also concerning lead, not to mention
what even Galen notes, that it will increase both in bulk
and weight if it be long kept in vaults or sellers, where
the air is gross and thick, as he collects from the swelling
of those pieces of lead that were imployed to fasten together the parts of old statues. Not to mention this, I
say, Boccacius Certaldus, as I find him quoted by a
diligent writer, has this passage touching the growth
of lead. " Fessularum mons (saies he) in Hetruria,
Florentiæ civitati imminens, lapides plumbarios habet;
qui si excidantur, brevi temporis spatio, novis incrementis
instaurantur; ut (annexes my author) tradit Boccacius
Certaldus, qui id compertissimum esse scribit. Nihil hoc
novi est; sed de eodem Plinius, lib. 34. *Hist. Natur.* cap.
17. dudum prodidit, inquiens, mirum in his solis plumbi
metallis, quod derelicta fertilius reviviscunt. In plum-
bariis secundo lapide ab amberga dictis ad asylum recre-
menta congesta in cumulos, exposita solibus pluviisque
paucis annis, reddunt suum metallum cum fœnore." I
might add to these (continues Carneades) many things
that I have met with concerning the generation of gold
and silver. But for fear of wanting time, I shall mention
but two or three narratives. The first you may find
recorded by Gerhardus the physick professor, in these
words. " In valle (saies he) Joachimica argentum
graminis modo et more è lapidibus mineræ velut è radice
excrevisse digiti longitudine, testis est Dr. Schreterus,
qui ejusmodi venas aspectu jucundas et admirabiles domi
suæ aliis sæpe monstravit et donavit. Item aqua cærulea
inventa est Annebergæ, ubi argentum erat adhuc in
primo ente, quæ coagulata redacta est in calcem fixi et
boni argenti."

The other two relations I have not met with in Latine
authors, and yet they are both very memorable in themselves, and pertinent to our present purpose.

The first I meet with in the commentary of Johannes
Valehius upon the *Kleine Baur*, in which that industrious
chymist relates, with many circumstances, that at a mine-

town (if I may so English the German *Bergstat*) eight
miles or leagues distant from Strasburg called Mariakirch,
a workman came to the overseer, and desired employment;
but he telling him that there was not any of the best sort
at present for him, added that till he could be preferred
to some such, he might in the meantime, to avoid idle-
ness, work in a grove or mine-pit thereabouts, which at
that time was little esteemed. This workman after some
weeks labour, had by a crack appearing in the stone upon
a stroak given near the wall, an invitation given him to
work his way through, which as soon as he had done,
his eyes were saluted by a mighty stone or lump which
stood in the middle of the cleft (that had a hollow place
behind it) upright, and in shew like an armed-man; but
consisted of pure fine silver having no vein or ore by it,
or any other additament, but stood there free, having
only underfoot something like a burnt matter; and yet
this one lump held in weight above a 1000 marks, which,
according to the Dutch account, makes 500 pound weight
of fine silver. From which and other circumstances my
author gathers; that by the warmth of the place, the noble
metalline spirits, (sulphureous and mercurial) were carried
from the neighbouring galleries or vaults, through other
smaller cracks and clefts into that cavity, and there
collected as in a close chamber or cellar; whereinto when
they were gotten, they did in process of time settle into
the forementioned precious mass of metal.

The other Germane relation is of that great traveller
and laborious chymist Johannes (not Georgius) Agricola;
who in his notes upon what Poppius has written of
antimony, relates, that when he was among the Hungarian
mines in the deep groves, he observed that there would
often arise in them a warm steam, (not of that malignant
sort which the Germans call *Shwadt*, which (saies he) is
a meer poyson, and often suffocates the diggers) which
fastened itself to the walls; and that coming again to
review it after a couple of dayes, he discerned that it was
all very fast, and glistering; whereupon having collected
it and distilled it *per retortam*, he obtained from it a fine

spirit: adding, that the mine-men informed him, that this steam, or damp (as the English men also call it, retaining the Dutch term) would at last have become a metal, as gold or silver.

I referr (saies Carneades) to another occasion, the use that may be made of these narratives towards the explicating the nature of metalls; and that of fixtness, malleableness, and some other qualities conspicuous in them. And in the meantime, this I may at present deduce from these observations; That 'tis not very probable, that, whensoever a mineral, or even a metal, is to be generated in the bowels of the earth, nature needs to have at hand both salt, and sulphur, and mercury to compound it of; for, not to urge that the two last relations seem less to favour the chymists than Aristotle, who would have metals generated of certain *halitus* or steams, the forementioned observations together, make it seem more likely that the mineral earths or those metalline steams (wherewith probably such earths are plentifully imbued) do contain in them some seminal rudiment, or something equivalent thereunto; by whose plastick power the rest of the matter, though perhaps terrestrial and heavy, is in tract of time fashioned into this or that metalline ore; almost (as I formerly noted) as that fair water was by the seminal principle of mint, pompions, and other vegetables, contrived into bodies answerable to such seeds. And that such alterations of terrestrial matter are not impossible, seems evident from that notable practice of the boylers of salt-petre, who unanimously observe, as well here in England as in other countries, that if an earth pregnant with nitre be deprived, by the affusion of water, of all its true and dissoluble salt, yet the earth will after some years yeeld them salt-petre again; for which reason some of the eminent and skilfullest of them keep it in heaps as a perpetual mine of salt-petre; whence it may appear, that the seminal principle of nitre latent in the earth does by degrees transforme the neighbouring matter into a nitrous body; for though I deny that some volatile nitre may by such earths be attracted (as they speak) out of the air, yet that the innermost parts of such great heaps that lye so remote from the air should borrow from it all the nitre they abound with, is not probable, for other reasons besides the remoteness of the air, though I have not the leasure to mention them.

And I remember, that a person of great credit, and well acquainted with the wayes of making vitriol, affirmed to me, that he had observed, that a kind of mineral which abounds in that salt, being kept within doors and not exposed (as is usual) to the free air and rains, did of itself in no very long time turn into vitriol, not only in the outward or superficial, but even in the internal and most central parts.

And I also remember, that I met with a certain kind of marchasite that lay together in great quantities under ground, which did, even in my chamber, in so few hours begin of itself to turne into vitriol, that we need not distrust the newly recited narrative. But to return to what I was saying of nitre; as nature made this salt-petre out of the once almost an inodorous earth it was bread in and did not find a very stinking and corrosive acid liquor, and a sharp alcalizate salt to compound it of, though these be the bodies into which the fire dissolves it; so it were not necessary that nature should make up all metals and other minerals of pre-existent salt, and sulphur, and mercury, though such bodies might by fire be obtained from it. Which one consideration duly weighed is very considerable in the present controversy: and to this agree well the relations of our two German chymists; for besides that it cannot be convincingly proved, it is not so much as likely that so languid and moderate a heat as that within the mines, should carry up to so great a height, though in the forme of fumes, salt, sulphur, and mercury; since we find in our distillations, that it requires a considerable degree of fire to raise so much as to the height of one foot not only salt, but even mercury itself, in close vessels. And if it be objected, that it seems by the stink that is sometimes observed when lightning falls down here below, that sulphureous steams may ascend very high without any extraordinary degree of heat; it may be answered, among other things, that the sulphur of silver is by chymists said to be a fixt sulphur, though not altogether so well digested as that of gold.

But, (proceeds Carneades) if it had not been to afford you some hints concerning the origine of metals, I need not have deduced anything from these observations; it not being necessary to the validity of my argument that my deductions from them should be irrefragable, because my adversaries the Aristotelians and vulgar chymists do not, I presume, know any better than I, *a priori*, of what ingredients nature compounds metals and minerals. For their argument to prove that those bodies are made up of such principles, is drawn *a posteriori;* I mean from this, that upon the analysis of mineral bodies they are resolved into those differing substances. That we may therefore examine this argument, let us proceed to consider what can be alledged in behalf of the elements from the resolutions of bodies by the fire; which you remember was the second topick whence I told you the arguments of my adversaries were desumed.

And that I may first dispatch what I have to say concerning minerals, I will begin the remaining part of my discourse with considering how the fire divides them.

And first, I have partly noted above, that though chymists pretend from some to draw salt, from others running mercury, and from others a sulphur; yet they have not hitherto taught us by any way in use among them to separate any one principle, whether salt, sulphur, or mercury, from all sorts of minerals without exception. And thence I may be allowed to conclude that there is not any of the elements that is an ingredient of all bodies, since there are some of which it is not so.

In the next place, supposing that either sulphur or mercury were obtainable from all sorts of minerals. Yet still this sulphur or mercury would be but acompounded, not an elementary body, as I told you already on another occasion. And certainly he that takes notice of the wonderful operations of quicksilver, whether it be common, or drawn from mineral bodies, can scarce be so inconsiderate as to think it of the very same nature with that immature and fugitive substance which in vegetables and animals chymists have been pleased to call their mercury. So that when mercury is got by the help of the fire out of a metal or other mineral body, if we will not suppose that it was not pre-existent in it, but produced by the action of the fire upon the concrete, we may at least suppose this quicksilver to have been a perfect body of its own kind (though perhaps less heterogeneous than more secondary mixts) which happened to be mingled *per minima*, and coagulated with the other substances, whereof the metal or mineral consisted. As may be exemplyfied partly by native vermilion wherein the quicksilver and sulphur being exquisitely blended both with one another, and that other course mineral stuff (whatever it be) that harbours them, make up a red body differing enough from both; and yet from which part of the quicksilver, and of the sulphur, may be easily enough obtained; partly by those mines wherein nature has so curiously incorporated silver with lead, that 'tis extremely difficult, and yet possible, to separate the former out of the latter; and partly too by native vitriol, wherein the metalline corpuscles are by skill and industry separable from the saline ones, though they be so con-coagulated with them, that the whole concrete is reckoned among salts.

And here I further observe, that I never could see any earth or water, properly so called, separated from either gold or silver (to name now no other metalline bodies) and therefore to retort the argument upon my adversaries, I may conclude, that since there are some bodies in which, for ought appears, there is neither earth nor water; I may be allowed to conclude, that neither of those two is an universal ingredient of all those bodies that are counted perfectly mixt, which I desire you would remember against anon.

It may indeed be objected, that the reason why from gold or silver we cannot separate any moisture, is, because that when it is melted out of the oar, the vehement fire requisite to its fusion forced away all the aqueous and fugitive moisture; and the like fire may do from the materials of glass. To which I shall answer, that I

remember I read not long since in the learned Josephus
Acosta, who relates it upon his own observation; that in
America (where he long lived) there is a kind of silver
which the Indians call *papas*, and sometimes (saies he)
they find pieces very fine and pure like to small round
roots, the which is rare in that metal, but usual in gold;
concerning which metal he tells us, that besides this they
find some which they call gold in grains, which he tells us
are small morsells of gold that they find whole without
mixture of any other metal, which hath no need of melting
or refining in the fire.

I remember that a very skilful and credible person
affirmed to me, that being in the Hungarian mines he had
the good fortune to see a mineral that was there digged
up, wherein pieces of gold of the length, and also almost
of the bigness of a humane finger, grew in the oar, as if
they had been parts and branches of trees.

And I have myself seen a lump of whitish mineral, that
was brought as a rarity to a great and knowing prince,
wherein there grew here and there in the stone, which
looked like a kind of sparr, divers little lumps of fine gold,
(for such I was assured that tryal had manifested it to be)
some of them seeming to be about the bigness of pease.

But that is nothing to what our Acosta subjoynes, which
is indeed very memorable, namely, that of the morsels
of native and pure gold, which we lately heard him mentioning, he had now and then seen some that weighed
many pounds; to which I shall add, that I myself have
seen a lump of oar not long since digged up, in whose
stony part there grew, almost like trees, divers parcels
though not of gold, yet of (what perhaps mineralists will
more wonder at) another metal which seemed to be very
pure or unmixt with any heterogeneous substances, and
were some of them as big as my finger, if not bigger.   But
upon observations of this kind, though perhaps I could,
yet I must not at present, dwell any longer.

To proceed therefore now (saies Carneades) to the consideration of the analysis of vegetables, although my
tryals give me no cause to doubt but that out of most of
them five differing substances may be obtained by the

fire, yet I think it will not be so easily demonstrated that these deserve to be called elements in the notion above explained.

And before I descend to particulars, I shall repeat and premise this general consideration, that these differing substances that are called elements or principles, differ not from each other as metals, plants and animals, or as such creatures as are immediately produced each by its peculiar seed, and constitutes a distinct propagable sort of creatures in the universe; but these are only various schemes of matter or substances that differ from each other, but in consistence (as running mercury and the same metal congealed by the vapor of lead) and some very few other accidents, as taste, or smell, or inflamability, or the want of them. So that by a change of texture not impossible to be wrought by the fire and other agents that have the faculty, not only to dissociate the small parts of bodies, but afterwards to connect them after a new manner, the same parcel of matter may acquire or lose such accidents as may suffice to denominate it salt, or sulphur, or earth. If I were fully to clear to you my apprehensions concerning this matter, I should perhaps be obliged to acquaint you with divers of the conjectures (for I must yet call them no more) I have had concerning the principles of things purely corporeal: for though because I seem not satisfied with the vulgar doctrines, either of the peripatetick or Paracelsian schooles, many of those that know me, (and perhaps, among them, Eleutherius himself) have thought me wedded to the Epicurean hypothesis, (as others have mistaken me for an Helmontian) yet if you knew how little conversant I have been with Epicurean authors, and how great a part of Lucretius himself I never yet had the curiosity to read, you would perchance be of another mind; especially if I were to entertain you at large, I say not, with my present notions; but with my former thoughts concerning the principles of things. But, as I said above, fully to clear my apprehensions would require a longer discourse than we can now have.

For, I should tell you that I have sometimes thought

it not unfit, that to the principles which may be assigned
to things, as the world is now constituted, we should, if we
consider the great mass of matter as it was whilst the
universe was in making, add another, which may conveniently enough be called an architectonick principle
or power; by which I mean those various determinations,
and that skilfull guidance of the motions of the small
parts of the universal matter by the most wise Author of
things, which were necessary at the beginning to turn
that confused chaos into this orderly and beautiful world;
and especially, to contrive the bodies of animals and
plants, and the seeds of those things whose kinds were
to be propagated.   For I confess I cannot well conceive,
how from matter, barely put into motion, and then left
to itself, there could emerge such curious fabricks as the
bodies of men and perfect animals, and such yet more
admirably contrived parcels of matter, as the seeds of
living creatures.

I should likewise tell you upon what grounds, and in
what sence, I suspected the principles of the world, as it
now is, to be three, *matter*, *motion*, and *rest*.   I say, *as
the world now is*, because the present fabrick of the
universe, and especially the seeds of things, together with
the establisht course of nature, is a requisite or condition,
upon whose account divers things may be made out by
our three principles, which otherwise would be very hard,
if possible, to explicate.

I should moreover declare in general (for I pretend
not to be able to do it otherwise) not only why I conceive
that colours, odours, tastes, fluidness and solidity, and
those other qualities that diversifie and denominate bodies
may intelligibly be deduced from these three; *but how two
of the three* Epicurean principles (which, I need not tell
you, are magnitude, figure, and weight) are themselves
deducible from matter and motion; since the latter of
these variously agitating, and, as it were, distracting the
former, must needs disjoyne its parts; which being
actually separated must each of them necessarily both
be of some size, and obtain some shape or other.   Nor
did I add to our principles the Aristotelian *privation*,

partly for other reasons, which I must not now stay to
insist on; and partly because it seems to be rather an
antecedent, or a *terminus à quo*, than a true principle,
as the starting-post is none of the horses legs or limbs.

I should also explain why and how I made rest, to be,
though not so considerable a principle of things, as motion;
yet a principle of them; partly because it is (for ought we
know) as ancient at least as it, and depends not upon
motion, nor any other quality of matter; and partly,
because it may enable the body in which it happens to be,
both to continue in a state of rest till some external force
put it out of that state, and to concur to the production
of divers changes in the bodies that hit against it, by
either quite stopping or lessening their motion (whilst the
body formerly at rest receives all or part of it into itself)
or else by giving a new byass, or some other modification,
to motion, that is, to the grand and primary instrument
whereby nature produces all the changes and other
qualities that are to be met with in the world.

I should likewise, after all this, explain to you how,
although matter, motion and rest, seemed to me to be
the catholick principles of the universe, I thought the
principles of particular bodies might be commodiously
enough reduced to two, namely *matter*, and (what com-
prehends the two other, and their effects) the result, or
aggregate, or complex of those accidents, which are the
motion or rest, (for in some bodies both are not to be
found) the bigness, figure, texture, and the thence resulting
qualities of the small parts, which are necessary to intitle
the body whereto they belong to this or that peculiar
denomination; and discriminating it from others to appro-
priate it to a determinate kind of things, (as yellowness,
fixtness, such a degree of weight, and of ductility, do
make the portion of matter wherein they concur, to be
reckoned among perfect metals, and obtain the name of
gold) this aggregate or result of accidents you may if you
please, call either *structure*, or texture, (though indeed,
that do not so properly comprehend the motion of the
constituent parts especially in case some of them be fluid)
or what other appellation shall appear most expressive.

Or if, retaining the vulgar terme, you will call it the *forme* of the thing it denominates, I shall not much oppose it; provided the word be interpreted to mean but what I have expressed, and not a scholastick *substantial forme*, which so many intelligent men profess to be to them altogether unintelligible.

But, (saies Carneades) if you remember that 'tis a sceptick speaks to you, and that 'tis not so much my present talk to make assertions as to suggest doubts, I hope you will look upon what I have proposed, rather as a narrative of my former conjectures touching the principles of things, than as a resolute declaration of my present opinions of them; especially since although they cannot but appear very much to their disadvantage, if you consider them as they are proposed without those reasons and explanations by which I could perhaps make them appear much less extravagant; yet I want time to offer you what may be alledged to clear and countenance these notions; my design in mentioning them unto you at present being, partly, to bring some light and confirmation to divers passages of my discourse to you; *partly* to shew you, that I do not (as you seem to have suspected) embrace all Epicurus his principles; but dissent from him in some main things, as well as from Aristotle and the chymists, in others; and *partly* also, or rather chiefly, to intimate to you the grounds upon which I likewise differ from Helmont in this, that whereas he ascribes almost all things, and even diseases themselves, to their determinate seeds; I am of opinion, that besides the peculiar fabricks of the bodies of plants and animals (and perhaps also of some metals and minerals) which I take to be effects of seminal principles, there are many other bodies in nature which have and deserve distinct and proper names, but yet do but result from such contextures of the matter they are made of, as may without determinate seeds be effected by heat, cold, artificial mixtures and compositions, and divers other causes which sometimes nature imployes of her own accord; and oftentimes man by his power and skill makes use of to fashion the matter according to his intentions. This may be exemplified both in the productions of nature, and in those of art; of the first sort I might name multitudes; but to shew how slight a variation of textures without addition of new ingredients may procure a parcel of matter divers names, and make it be lookt upon as different things;

I shall invite you to observe with me, that clouds, rain, hail, snow, frost, and ice, may be but water, having its parts varied as to their size and distance in respect of each other, and as to motion and rest. And among artificial productions we may take notice (to skip the chrystals of tartar) of glass, regulus martis stellatus, and particularly of the sugar of lead, which though made of that insipid metal and sowre salt of vinegar, has in it a sweetness surpassing that of common sugar, and divers other qualities, which being not to be found in either of its two ingredients, must be confessed to belong to the concrete itself, upon the account of its texture.

This consideration premised, it will be, I hope, the more easie to perswade you that the fire may as well produce some new textures in a parcel of matter, as destroy the old.

Wherefore hoping that you have not forgot the arguments formerly imployed against the doctrine of the *tria prima*; namely that the salt, sulphur, and mercury, into which the fire seems to resolve vegetable and animal bodies, are yet compounded, not simple and elementary substances; and that (as appeared by the experiment of pompions) the *tria prima* may be made out of water; hoping I say, that you remember these and the other things that I formerly represented to the same purpose, I shall now add only, that if we doubt not the truth of some of Helmont's relations, we may well doubt whether any of these heterogeneities be (I say not pre-existent, so as to convene together, when a plant or animal is to be constituted, but) so much as inexistent in the concrete whence they are obtained, when the chymist first goes about to resolve it; for, not to insist upon the uninflamable spirit of such concretes, because that may be pretended to be but a mixture of phlegme and salt; the oyle or sulphur of vegetables or animals is, according to him, reducible by the help of lixiviate salts into sope; as that sope is by the help of repeated distillations from a *caput mortuum* of chalk into insipid water. And as for the saline substance that seems separable from mixt bodies; the same Helmont's tryals give us cause to think, that it may be a production of the fire which by transporting and otherwise altering the particles of the matter, does bring it to a saline nature.

For I know (saies he, in the place formerly alledged to another purpose) a way to reduce all stones into a mere salt of equal weight with the stone whence it was produced, and that without any of the least either sulphur or mercury; which asseveration of my author would perhaps seem less incredible to you, if I durst acquaint you with all I could say upon that subject. And hence by the way you may also conclude that the sulphur and mercury, as they call them, that chymists are wont to obtain from compound bodies by the fire, may possibly in many cases be the productions of it; since if the same bodies had been wrought upon by the agents employed by Helmont, they would have yielded neither sulphur nor mercury; and those portions of them, which the fire would have presented us in the forme of sulphureous and mercurial bodies, would have, by Helmont's method, been exhibited to us in the form of salt.

But though (saies Eleutherius) you have alledged very plausible arguments against the *tria prima*, yet I see not how it will be possible for you to avoid acknowledging that earth and water are elementary ingredients, though not of mineral concretes, yet of all animal and vegetable bodies; since if any of these of what sort soever be committed to distillation, there is regularly and constantly separated from it a phlegme or aqueous part, and a *caput mortuum* or earth.

I readily acknowledge (answers Carneades) it is not so easy to reject water and earth (and especially the former) as 'tis to reject the *tria prima*, from being the elements of mixt bodies; but 'tis not every difficult thing that is impossible.

I consider then, as to water, that the chief qualities

which make men give that name to any visible substance, are that it is fluid or liquid, and that it is insipid and inodorous. Now as for the taste of these qualities, I think you have never seen any of those separated substances that the chymists call phlegme which was perfectly devoid both of taste and smell: and if you object, that yet it may be reasonably supposed, that since the whole body is liquid, the mass is nothing but elementary water faintly imbued with some of the saline or sulphureous parts of the same concrete, which it retained with it upon its separation from the other ingredients. To this I answer, that this objection would not appear so strong as it is plausible, if chymists understood the nature of fluidity and compactness; and that, as I formerly observed, to a bodies being fluid there is nothing necessary, but that it be divided into parts small enough; and that these parts be put into such a motion among themselves as to glide some this way and some that way, along each other's surfaces. So that although a concrete were never so dry, and had not any water or other liquor inexistent in it, yet such a comminution of its parts may be made, by the fire or other agents, as to turn a great portion of them into liquor. Of this truth I will give an instance, employed by our friend here present as one of the most conducive of his experiments to illustrate the nature of salts. If you take then sea salt, and melt it in the fire to free it from the aqueous parts, and afterwards distill it with a vehement fire from burnt clay, or any other, as dry a *caput mortuum* as you please, you will, as chymists confess by teaching it, drive over a good part of the salt in the form of a liquor. And to satisfy some ingenious men, that a great part of this liquor was still true sea salt brought by the operation of the fire into corpuscles so small, and perhaps so advantageously shaped, as to be capable of the forme of a fluid body, he did in my presence poure to such spiritual salts a due proportion of the spirit (or salt and phlegme) of urine, whereby having evaporated the superfluous moisture, he soon obtained such another concrete, both as to taste and smell, and easie sublimableness as common salt armoniack, which you know is made up of gross and undistilled sea salt united with the salts or urine and of soot, which two are very near of kin to each other. And further, to manifest that the corpuscles of sea salt and the saline ones of urine retain their several natures in this concrete, he mixt it with a convenient quantity of salt of tartar, and committing it to distillation soon regained his spirit of urine in a liquid form by itself, the sea salt staying behind with the salt of tartar. Wherefore it is very possible that dry bodies may by the fire be reduced to liquors without any separation of elements, but barely by a certain kind of dissipation and comminution of the matter, whereby its parts are brought into a new state. And if it be still objected, that the phlegme of mixt bodies must be reputed water, because so weak a taste needs but a very small proportion of salt to impart it; it may be replyed, that for ought appears, common salt and divers other bodies, though they be distilled never so dry, and in never so close vessels, will yeeld each of them pretty store of a liquor, wherein though (as I lately noted) saline corpuscles abound, yet there is besides a large proportion of phlegme, as may easily be discovered by coagulating the saline corpuscles with any convenient body; as I lately told you, our friend coagulated part of the spirit of salt with spirit of urine: and as I have divers times separated a salt from oyle of vitriol itself (though a very ponderous liquor and drawn from a saline body) by boyling it with a just quantity of mercury, and then washing the newly coagulated salt from the precipitate with fair water. Now to what can we more probably ascribe this plenty of aqueous substance afforded us by the distillation of such bodies, than unto this, that among the various operations of the fire upon the matter of a concrete divers particles of that matter are reduced to such a shape and bigness, as is requisite to compose such a liquor as chymists are wont to call phlegme or water. How I conjecture this change may be effected, 'tis neither necessary for me to tell you, nor possible to do so without a much longer discourse than were now seasonable. But I desire you would with me reflect upon what I formerly told you concerning the change of quicksilver into water; for that water having but a very faint taste, if any whit more than divers of those liquors that chymists referr to phlegme, by that experiment it seems evident, that even a metalline body, and therefore much more such as are but vegetable or animal, may by a simple operation of the fire be turned in great part into water. And since those I dispute with are not yet able out of gold, or silver, or divers other concretes to separate anything like water; I hope I may be allowed to conclude against them, that water itself is not an universal and pre-existent ingredient of mixt bodies.

But as for those chymists that, supposing with me the truth of what Helmont relates of the alkahest's wonderful effects, have a right to press me with his authority concerning them, and to alledge that he could transmute all reputed mixt bodies into insipid and mere water; to those I shall represent, that though his affirmations conclude strongly against the vulgar chymists (against whom I have not therefore scrupled to employ them) since they evince that the commonly reputed principles or ingredients of things are not permanent and indestructible, since they may be further reduced into insipid phlegme differing from them all; yet till we can be allowed to examine this liquor, I think it not unreasonable to doubt whether it be not something else than mere water. For I find not any other reason given by Helmont of his pronouncing it so, than that it is insipid. Now sapour being an accident or an affection of matter that relates to our tongue, palate and other organs of taste, it may very possibly be, that the small parts of a body may be of such a size and shape, as either by their extream littleness, or by their slenderness, or by their figure, to be unable to pierce into and make perceptible impression upon the nerves or membranous parts of the organs of taste, and yet may be fit to work otherwise upon divers other bodies than mere water can, and consequently to disclose itself to be of a nature farr enough from elementary. In silke dyed red or of any other colour, whilst many contiguous threads make up a skein, the colour of the silke is conspicuous; but if only a very few of them be lookt upon, the colour will appear much fainter than before. But if you take out one simple thread, you shall not easily be able to discern any colour at all; so subtile an object having not the force to make upon the optick nerve an impression great enough to be taken notice of. It is also observed, that the best sort of oyl-olive is almost tasteless, and yet I need not tell you how exceedingly distant in nature oyle is from water. The liquor into which I told you, upon the relation of Lully an eye-witness, that mercury might be transmuted, has sometimes but a very languid, if any taste; and yet its operations even upon some mineral bodies are very peculiar. Quicksilver itself also, though the corpuscles it consists of be so very small, as to get into the pores of that closest and compactest of bodies, gold, is yet (you know) altogether tasteless. And our Helmont several times tells us, that fair water, wherein a little quantity of quicksilver has lain for some time, though it acquire no certain taste or other sensible quality from the quicksilver; yet it has a power to destroy wormes in human bodies; which he does much, but not causelessly extoll. And I remember, a great lady, that had been eminent for her beauty in divers courts, confessed to me, that this insipid liquor was of all innocent washes for the face the best that she ever met with.

And here let me conclude my discourse, concerning such waters or liquors as I have hitherto been examining, with these two considerations. Whereof the first is, That by reason of our being wont to drink nothing but wine, bear, cider, or other strongly tasted liquors, there may be in several of those liquors, that are wont to pass for insipid phlegme, very peculiar and distinct tastes, though unheeded (and perhaps not to be perceived) by us. For to omit what naturalists affirm of apes, (and which probably may be true of divers other animals) that they have a more exquisite palate than men: among men themselves, those that are wont to drink nothing but water, may (as I have tryed in myself) discern very sensibly a great difference of tastes in several waters, which one unaccustomed to drink water would take to be all alike insipid. And this is the *first* of my two considerations. The other is, That it is not impossible that the corpuscles, into which a body is dissipated by the fire, may by the operation of the same fire have their figures so altered, or may be by associations with one another brought into little masses of such a size and shape, as not to be fit to make sensible impressions on the tongue. And that you may not think such alterations impossible, be pleased to consider with me, that not only the sharpest spirit of vinegar having dissolved as much corall as it can, will coagulate with it into a substance, which, though soluble in water like salt, is incomparably less strongly tasted than the vinegar was before; but (what is more considerable) though the acid salts that are carried up with quicksilver in the preparation of common sublimate are so sharp, that being moistened with water it will corrode some of the metals themselves; yet this corrosive sublimate being twice or thrice re-sublimed with a full proportion of insipid quicksilver, constitutes (as you know) that factitious concrete which the chymists call *mercurius dulcis ;* not because it is sweet, but because the sharpness of the corrosive salts is so taken away by their combination with the mercurial corpuscles, that the whole mixture when it is prepared is judged to be insipid.

And thus (continues Carneades) having given you some reasons why I refuse to admit elementary water for a constant ingredient of mixt bodies, it will be easie for me to give you an account why I also reject earth.

For first, it may well be suspected that many substances pass among chymists under the name of earth, because, like it, they are dry, and heavy, and fixt, which yet are very farr from an elementary nature. This you will not think improbable, if you recall to mind what I formerly told you concerning what chymists call the dead earth of things, and especially touching the copper to be drawn from the *caput mortuum* of vitriol; and if also you allow me to subjoyne a casual but memorable experiment made by Johannes Agricola upon the *terra damnata* of brimstone. Our author then tells us (in his notes upon

Popius) that in the year 1621 he made an oyle of sulphur; the remaining fæces he reverberated in a moderate fire fourteen dayes; afterwards he put them well luted up in a wind oven, and gave them a strong fire for six hours, purposing to calcine the fæces to a perfect whiteness, that he might make something else out of them. But coming to break the pot, he found above but very little fæces, and those grey and not white; but beneath there lay a fine red regulus which he first marvelled at and knew not what to make of, being well assured that not the least thing, besides the fæces of the sulphur, came into the pot; and that the sulphur itself had only been dissolved in linseed oyle; this regulus he found heavy and malleable almost as lead; having caused a goldsmith to draw him a wire of it, he found it to be of the fairest copper, and so rightly coloured, that a Jew of Prague offered him a great price for it. And of this metal he saies he had 12 *loth* (or six ounces) out of one pound of ashes or fæces. And this story may well incline us to suspect that since the *caput mortuum* of the sulphur was kept so long in the fire before it was found to be anything else than a *terra damnata*, there may be divers other residences of bodies which are wont to pass only for the terrestrial fæces of things, and therefore to be thrown away as soon as the distillation or calcination of the body that yeelded them is ended; which yet, if they were long and skilfully examined by the fire, would appear to be differing from elementary earth. And I have taken notice of the unwarrantable forwardness of common chymists to pronounce things useless fæces, by observing how often they reject the *caput mortuum* of verdegrease; which is yet so farr from deserving that name, that not only by strong fires and convenient additaments it may in some hours be reduced into copper, but with a certain flux powder I sometimes make for recreation, I have in two or three minutes obtained that metal from it. To which I may add, that having for tryall sake kept Venetian talck in no less a heat than that of a glass furnace, I found after all the brunt of the fire it had indured, the remaining body, though brittle and discoloured, had not lost very

much of its former bulke, and seemed still to be nearer
of kin to talck than to mere earth.   And I remember too,
that a candid mineralist, famous for his skill in trying of
oars, requesting me one day to procure him a certain
American mineral earth of a virtuoso, who he thought
would not refuse me; I enquired of him why he seemed
so greedy of it: he confessed to me that this gentleman
having brought that earth to the publick say-masters;
and they upon their being unable by any means to bring
it to fusion or make it fly away, he (the relator) had procured a little of it; and having tryed it with a peculiar
flux, separated from it near a third part of pure gold; so
great mistakes may be committed in hastily concluding
things to be useless earth.

Next, it may be supposed, that as in the resolution of
bodies by the fire some of the dissipated parts may, by
their various occursion occasioned by the heat, be brought
to stick together so closely as to constitute corpuscles
too heavy for the fire to carry away; the aggregate of
which corpuscles is wont to be called ashes or earth; so
other agents may resolve the concrete into minute parts
after so differing a manner, as not to produce any *caput
mortuum*, or dry and heavy body.   As you may remember
Helmont above informed us, that with his great dissolvent
he divided a coal into two liquid and volatile bodies,
æquiponderant to the coal, without any dry or fixt
residence at all.

And indeed, I see not why it should be necessary that
all agents that resolve bodies into portions of differing
qualified matter must work on them the same way, and
divide them into just such parts, both for nature and
number, as the fire dissipates them into.   For since,
(as I noted before) the bulk and shape of the small parts
of bodies, together with their fitness and unfitness to be
easily put into motion, may make the liquors or other
substances such corpuscles compose, as much to differ
from each other as do some of the chymical principles:
why may not something happen in this case, not unlike
what is usuall in the grosser divisions of bodies by mecha-
nical instruments?   Where we see that some tools reduce

wood, for instance, into parts of several shapes, bigness, and other qualities, as hatchets and wedges divide it into grosser parts; some more long and slender, as splinters; and some more thick and irregular, as chips; but all of considerable bulk; but files and saws make a comminution of it into dust; which, as all the others, is of the more solid sort of parts; whereas others divide it into long and broad, but thin and flexible parts, as do planes: and of this kind of parts itself there is also a variety according to the difference of the tools employed to work on the wood; the shavings made by the plane being in some things differing from those shives or thin and flexible pieces of wood that are obtained by borers, and these from some others obtainable by other tools. Some chymical examples applicable to this purpose I have elsewhere given you. To which I may add, that whereas, in a mixture of sulphur and salt of tartar well melted and incorporated together, the action of pure spirit of wine digested on it is to separate the sulphureous from the alcalizate parts, by dissolving the former and leaving the latter: the action of wine (probably upon the score of its copious phlegme) upon the same mixture is to divide it into corpuscles consisting of both alcalizate and sulphureous parts united. And if it be objected, that this is but a factitious concrete; I answer, that however the instance may serve to illustrate what I proposed, if not to prove it; and that nature herself doth in the bowels of the earth make decompounded bodies, as we see in vitriol, cinnaber, and even in sulphur itself; I will not urge that the fire divides new milk into five differing substances; but runnet and acid liquors divide it into a coagulated matter and a thin whey: and on the other side churning divides it into butter and buttermilk, which may either of them yet be reduced to other substances differing from the former. I will not press this, I say, nor other instances of this nature, because I cannot in few words answer what may be objected, that these concretes sequestred without the help of the fire may by it be further divided into hypostatical principles. But I will rather represent, that whereas the same spirit of wine will

dissociate the parts of camphire, and make them one
liquor with itself; *aqua fortis* will also disjoyne them, and
put them into motion; but so as to keep them together,
and yet alter their texture into the form of an oyle. I
know also an uncompounded liquor, that an extraordinary chymist would not allow to be so much as
saline, which doth (as I have tryed) from coral itself
(as fixt as divers judicious writers assert that concrete
to be) not only obtain a noble tincture without the intervention of nitre or other salts; but will carry over the
tincture in distillation. And if some reasons did not
forbid me, I could now tell you of a menstruum I make
myself, that doth more odly dissociate the parts of minerals
very fixt in the fire. So that it seems not incredible,
that there may be some agent or way of operation found,
whereby this or that concrete, if not all firme bodies, may
be resolved into parts so very minute and so apt to stick
close to one another, that none of them may be fixt enough
to stay behind in a strong fire, and to be incapable of
distillation; nor consequently to be looked upon as earth.
But to return to Helmont; the same author somewhere
supplys me with another argument against the earth's
being such an element as my adversaries would have it.
For he somewhere affirmes, that he can reduce all the
terrestrial parts of mixt bodies into insipid water; whence
we may argue against the earth's being one of their
elements, even from that notion of elements, which you
may remember Philoponus recited out of Aristotle himself, when he lately disputed for his chymists against
Themistius. And here we may on this occasion consider,
that since a body, from which the fire hath driven away
its looser parts, is wont to be looked upon as earth, upon
the account of its being endowed with both these qualities,
tastlesnesse and fixtnesse, (for salt of tartar, though fixt,
passes not among the chymists for earth, because 'tis
strongly tasted) if it be in the power of natural agents to
deprive the *caput mortuum* of a body of either of those
two qualities, or to give them both to a portion of matter
that had them not both before, the chymists will not
easily define what part of a resolved concrete is earth,

and make out, that that earth is a primary, simple, and indestructible body. Now there are some cases wherein the more skilful of the vulgar chymists themselves pretend to be able, by repeated cohobations and other fit operations, to make the distilled parts of a concrete bring its own *caput mortuum* over the helme, in the forme of a liquor, in which state being both fluid and volatile, you will easily believe it would not be taken for earth. And indeed by a skilful, but not vulgar, way of managing some concretes, there may be more effected in this kind, than you perhaps would easily think. And on the other side, that either earth may be generated, or at least bodies that did not before appear to be near totally earth, may be so altered as to pass for it, seems very possible, if Helmont have done that by art which he mentions in several places; especially where he saies that he knowes waies whereby sulphur once dissolved is all of it fixed into a terrestrial powder, and the whole body of salt-petre may be turned into earth: which last he elsewhere saies is done by the odour only of a certain sulphureous fire. And in another place he mentions one way of doing this, which I cannot give you an account of; because the materials I had prepared for trying it, were by a servant's mistake unhappily thrown away.

And these last arguments may be confirmed by the experiment I have often had occasion to mention concerning the mint I produced out of water. And partly by an observation of Rondeletius concerning the growth of animals also, nourished but by water, which I remembered not to mention, when I discoursed to you about the production of things out of water. This diligent writer then in his instructive book of fishes, affirmes that his wife kept a fish in a glass of water without any other food for three years; in which space it was constantly augmented, till at last it could not come out of the place at which it was put in, and at length was too big for the glass itself, though that were of a large capacity. And because there is no just reason to doubt, that this fish, if distilled would have yeelded the like differing substances with other animals; and however, because the mint, which I had out of water, afforded me upon distillation a good quantity of charcoal; I think I may from thence inferr, that earth itself may be produced out of water; or if you please, that water may be transmuted into earth; and consequently, that though it could be proved, that earth is an ingredient actually inexistent in the vegetable and animal bodies whence it may be obtained by fire: yet it would not necessarily follow, that earth, as a pre-existent element does with other principles convene to make up those bodies whence it seems to have been separated.

After all is said (saies Eleutherius) I have yet something to object, that I cannot but think considerable, since Carneades himself alledged it as such; for, (continues Eleutherius smiling) I must make bold to try whether you can as luckily answer your own arguments, as those of your antagonists, I mean (pursues he) that part of your concessions, wherein you cannot but remember, that you supplyed your adversaries with an example to prove that there may be elementary bodies, by taking notice that gold may be an ingredient in a multitude of differing mixtures, and yet retain its nature, notwithstanding all that the chymists by their fires and corrosive waters are able to do to destroy it.

I sufficiently intimated to you at that time (replies Carneades) that I proposed this example, chiefly to shew you how nature may be conceived to have made elements, not to prove that she actually has made any; and you know, that *a posse ad esse* the inference will not hold. But (continues Carneades) to answer more directly to the objection drawn from gold, I must tell you, that though I know very well that divers of the more sober chymists have complained of the vulgar chymists, as of mounte-banks or cheats, for pretending so vainly, as hitherto they have done, to destroy gold; yet I know a certain menstruum (which our friend has made, and intends shortly to communicate to the ingenious) of so piercing and powerful a quality, that if notwithstanding much care, and some skill, I did not much deceive myself, I have with it really destroyed even refined gold, and brought it into a metalline body of another colour and nature, as I found by tryals purposely made. And if some just considerations did not for the present forbid it, I could perchance here shew you by another experiment or two of my own trying, that such menstruums may be made as to entice away and retain divers parts from bodies, which even the more judicious and experienced spagyrists have pronounced irresoluble by the fire. Though (which I desire you would mark) in neither of these instances, the gold or precious stones be analyzed into any of the *tria prima*, but only reduced to new concretes. And indeed there is a great disparity betwixt the operations of the several agents whereby the parts of a body come to be dissipated. As if (for instance) you dissolve the purer sort of vitriol in common water, the liquor will swallow up the mineral, and so dissociate its corpuscles, that they will seem to make up but one liquor with those of the water; and yet each of these corpuscles retains its nature and texture, and remains a vitriolate and compounded body. But if the same vitriol be exposed to a strong fire, it will then be divided not only, as before, into smaller parts, but into heterogeneous substances, each of the vitriolate corpuscles that remained entire in the water, being itself upon the destruction of its former texture dissipated or divided into new particles of differing qualities. But instances more fitly applicable to this purpose I have already given you. Wherefore to return to what I told you about the destruction of gold; that experiment invites me to represent to you, that though there were either saline, or sulphureous, or terrestrial portions of matter, whose parts were so small, so firmly united together, or of a figure so fit to make them cohere to one another, (as we see that in quicksilver broken into little globes, the parts brought to touch one another do immediately reimbody) that neither the fire, nor the usual agents, employed by chymists, are piercing enough to divide their parts, so as to destroy the texture of the single corpuscles; yet it would not necessarily follow, that such permanent bodies were elementary; since 'tis possible there may be agents found in nature, some of whose parts may be of such a size and figure as to take better hold of some parts of these seemingly elementary corpuscles than these parts do of the rest, and consequently may carry away such parts with them, and so dissolve the texture of the corpuscle by pulling its parts asunder. And if it be said, that at least we may this way discover the elementary ingredients of things by observing into what substances these corpuscles, that were reputed pure are divided; I answer, that 'tis not necessary that such a discovery should be practicable. For if the particles of the dissolvent do take such firm hold of those of the dissolved body, they must constitute together new bodies, as well as destroy the old; and the strickt union, which according to this hypothesis may well be supposed betwixt the parts of the emergent body, will make it as little to be expected that they should be pulled asunder, but by little parts of matter, that to divide them associate themselves and stick extremely close to those of them which they sever from their former adherents, besides that it is not impossible, that a corpuscle supposed to be elementary may have its nature changed, without suffering a divorce of its parts, barely by a new texture effected by some powerful agent; as I formerly told you, the same portion of matter may easily by the operation of the fire be turned at pleasure into the form of a brittle and transparent, or an opacous and malleable body.

And indeed, if you consider how farr the bare change of texture, whether made by art or nature (or rather by nature with or without the assistance of man) can go in producing such new qualities in the same parcel of matter, and how many inanimate bodies (such as are all the chymical productions of the fire) we know are denominated and distinguished not so much by any imaginary substantial form, as by the aggregate of these qualities; if you consider these things, I say, and that the varying of either figure, or the size, or the motion, or the situation, or connexion of the corpuscles whereof any of these bodies is composed, may alter the fabrick of it, you will possibly be invited to suspect with me, that there is no great need that nature should alwaies have elements beforehand, whereof to make such bodies as we call mixts. And that it is not so easie as chymists and others have hitherto imagined, to discern, among the many differing substances that may without any extraordinary skill be obtained from the same portion of matter, which ought to be esteemed exclusively to all the rest, its inexistent elementary ingredients; much less to determine what primogeneal and simple bodies convened together to compose it. To exemplify this, I shall add to what I have already on several occasions represented, but this single instance.

You may remember (Eleutherius) that I formerly intimated to you, that besides mint and pompions, I produced divers other vegetables of very differing natures out of water. Wherefore you will not, I presume, think it incongruous to suppose, that when a slender vine-slip is set into the ground, and takes root there, it may likewise receive its nutriment from the water attracted out of the earth by its roots, or impelled by the warmth of the sun, or pressure of the ambient air into the pores of them. And this you will the more easily believe, if you ever observed what a strange quantity of water will drop out of a wound given to the vine, in a convenient place, at a seasonable time in the spring; and how little of taste or smell this *aqua vitis*, as physitians call it, is endowed with, notwithstanding what concoction or alteration it may receive in its passage through the vine, to discriminate it from common water. Supposing then this liquor, at its first entrance into the roots of the vine, to be common water; let us a little consider how many various substances may be obtained from it; though to do so, I must repeat somewhat that I had a former occasion to touch upon. And first, this liquor being digested in the plant, and assimilated by the several parts of it, is turned into the wood, bark, pith, leaves, etc. of the vine; the same liquor may be further dryed, and fashioned into vine-buds, and these a while after are advanced unto sowre grapes, which expressed yeeld verjuice, a liquor very differing in several qualities both from wine and other liquors obtainable from the vine: these sowre grapes, being by the heat of the sun concocted and ripened, turne to well tasted grapes; these, if dryed in the sun and distilled, afford a fætid oyle and a piercing empyreumatical spirit, but not a vinous spirit; these dryed grapes or raisins, boyled in a convenient proportion of water, make a sweet liquor, which, being betimes distilled, afford an oyle and spirit much like those of the raisins themselves; if the juice of the grapes be squeezed out and put to ferment, it first becomes a sweet and turbid liquor, then grows lesse sweet and more clear, and then affords in common distillations not an oyle but a spirit, which, though inflamable like oyle, differs much from it, in that it is not fat, and that it will readily mingle with water. I have likewise without addition obtained in processe of time (and by an easie way which I am ready to teach you) from one of the noblest sorts of wine, pretty store of pure and curiously figured chrystals of salt, together with a great proportion of a liquor as sweet almost as honey; and these I obtained not from must, but true and sprightly wine; besides the vinous liquor, the fermented juice of grapes is partly turned into liquid dregs or leeze, and partly into that crust or dry feculancy that is commonly called tartar; and this tartar may by the fire be easily divided into five differing substances; four of which are not acid, and the other not so manifestly acid as the tartar itself; the same vinous juice after some time, especially if it be not carefully kept, degenerates into that very sowre liquor called vinegar; from which you may obtain by the fire a spirit and a chrystalline salt differing enough from the spirit and lixiviate salt of tartar. And if you poure the dephlegmed spirit of the vinegar upon the salt of tartar, there will be produced such a conflict or ebullition, as if there were scarce two more contrary bodies in nature; and oftentimes in this vinegar you may observe part of the matter to be turned into an innumerable company of swimming animals, which our friend having divers years ago observed, hath in one of his papers taught us how to discover clearly without the help of a microscope.

Into all these various schemes of matter, or differingly

qualifyed bodies, besides divers others that I purposely forbear to mention, may the water, that is imbibed by the roots of the vine, be brought, partly by the formative power of the plant, and partly by supervenient agents or causes, without the visible concurrence of any extraneous ingredient; but if we be allowed to add to the productions of this transmuted water a few other substances, we may much encrease the variety of such bodies; although in this second sort of productions, the vinous parts seem scarce to retain anything of the much more fixed bodies wherewith they were mingled, but only to have by their mixture with them acquired such a disposition, that in their recess occasioned by the fire they came to be altered as to shape, or bigness, or both, and associated after a new manner. Thus, as I formerly told you, I did by the addition of a *caput mortuum* of antimony, and some other bodies unfit for distillation, obtain from crude tartar, store of a very volatile and chrystalline salt, differing very much in smell and other qualities from the usuall salts of tartar.

But (saies Eleutherius, interrupting him at these words) if you have no restraint upon you, I would very gladly before you go any further, be more particularly informed, how you make this volatile salt, because (you know) that such multitudes of chymists have by a scarce imaginable variety of waies, attempted in vain the volatilization of the salt of tartar, that divers learned spagyrists speak as if it were impossible to make anything out of tartar, that shall be volatile in a saline forme, or, as some of them express it, *in forma sicca*. I am very farr from thinking (answers Carneades) that the salt I have mentioned is that which Paracelsus and Helmont mean, when they speak of *sal tartari volatile,* and ascribe such great things to it. For the salt I speak of falls extremely short of those vertues, not seeming in its taste, smel, and other obvious qualities, to differ very much (though something it does differ) from salt of hartshorn, and other volatile salts drawn from the distilled parts of animals. Nor have I yet made tryals enough to be sure, that it is a pure salt of tartar without participating anything at all of the

nitre, or antimony. But because it seems more likely
to proceed from the tartar, than from any of the other
ingredients, and because the experiment is in itself not
ignoble, and luciferous enough (as shewing a new way to
produce a volatile salt, contrary to acid salts, from bodies
that otherwise are observed to yeeld no such liquor, but
either only, or chiefly, acid ones,) I shall, to satisfie you,
acquaint you before any of my other friends with the
way I now use (for I have formerly used some others)
to make it.

Take then of good antimony, salt-petre and tartar, of
each an equal weight, and of quicklime halfe the weight
of any one of them; let these be powdered and well
mingled; this done, you must have in readiness a long
neck or retort of earth, which must be placed in a furnace
for a naked fire, and have at the top of it a hole of a convenient bigness, at which you may cast in the mixture,
and presently stop it up again; this vessel being fitted
with a large receiver must have fire made under it, till
the bottom of the sides be red hot, and then you must
cast in the above prepared mixture, by about half a
spoonful (more or less) at a time, at the hole made for
that purpose; which being nimbly stopt, the fumes will
pass into the receiver and condense there into a liquor,
that being rectified will be of a pure golden colour, and
carry up that colour to a great height; this spirit abounds
in the salt I told you of, part of which may easily enough
be separated by the way I use in such cases, which is,
to put the liquor into a glass egg, or bolthead with a long
and narrow neck. For if this be placed a little inclining
in hot sand, there will sublime up a fine salt, which, as
I told you, I find to be much of kin to the volatile salts
of animals: for like them it has a saltish, not an acid
salt; it hisses upon the affusion of spirit of nitre, or oyle
of vitriol; it precipitates corals dissolved in spirit of
vinegar; it turnes the blew syrup of violets immediately
green; it presently turnes the solution of sublimate into
a milkie whiteness; and in summ, has divers operations
like those that I have observed in that sort of salts to
which I have resembled it: and is so volatile, that for

distinction sake, I call it *sal tartari fugitivus*. What vertues it may have in physick I have not yet had the opportunity to try; but I am apt to think they will not be despicable. And besides that, a very ingenious friend of mine tells me he hath done great matters against the stone with a preparation not very much differing from ours: a very experienced Germane chymist finding that I was unacquainted with the waies of making this salt, told me that in a great city in his country, a noted chymist prizes it so highly, that he had a while since procured a priviledge from the magistrates, that none but he, or by his licence, should vent a spirit made almost after the same way with mine, save that he leaves out one of the ingredients, namely the quicklime. But, (continues Carneades) to resume my former discourse where your curiosity interrupted it;

Tis also a common practice in France to bury thin plates of copper in the marc (as the French call it) or husks of grapes, whence the juice has been squeezed out in the wine-press; and by this means the more saline parts of those husks, working by little and little upon the copper, coagulate themselves with it into that blewish green substance we in English call verdigrease. Of which I therefore take notice, because having distilled it in a naked fire, I found, as I expected, that by the association of the saline with the metalline parts, the former were so altered, that the distilled liquor, even without rectification, seemed by smell and taste, strong almost like *aqua fortis*, and very much surpassed the purest and most rectified spirit of vinegar that ever I made. And this spirit I therefore ascribe to the salt of the husks altered by their co-mixture with the copper (though the fire afterwards divorce and transmute them) because I found this latter in the bottom of the retort in the forme of a *crocus* or reddish powder: and because copper is of too sluggish a nature to be forced over in close vessels by no stronger a heat. And that which is also somewhat remarkable in the distillation of good verdigrease, (or at least of that sort that I used) is this, that I never could observe that it yeelded me any oyl, (unless

a little black slime which was separated in rectification
may pass for oyle) though both tartar and vinegar (especially the former) will by distillation yeeld a moderate
proportion of it.   If likewise you poure spirit of vinegar
upon calcined lead, the acid salt of the liquor will by
its commixture with the metalline parts, though insipid,
acquire in few hours a more than saccharine sweetness;
and these saline parts being by a strong fire distilled from
the lead wherewith they were imbodyed, will, as I formerly
also noted to a different purpose, leave the metal behind
them altered in some qualities from what it was, and will
themselves ascend, partly in the form of an unctuous
body or oyle, partly in that of phlegme, but for the greatest
part in the forme of a subtile spirit, indowed, besides
divers new qualities which I am not now willing to take
notice of, with a strong smell very much other than that
of vinegar, and a piercing taste quite differing both from
the sowreness of the spirit of vinegar, and the sweetness
of the sugar of lead.

To be short, as the difference of bodies may depend
merely upon that of the schemes whereinto their common
matter is put; so the seeds of things, the fire and the
other agents are able to alter the minute parts of a body
(either by breaking them into smaller ones of differing
shapes, or by uniting together these fragments with the
unbroken corpuscles, or such corpuscles among themselves) and the same agents partly by altering the shape
or bigness of the constituent corpuscles of a body, partly
by driving away some of them, partly by blending others
with them, and partly by some new manner of connecting
them, may give the whole portion of matter a new texture
of its minute parts, and thereby make it deserve a new
and distinct name.   So that according as the small parts
of matter recede from each other, or work upon each other,
or are connected together after this or that determinate
manner, a body of this or that denomination is produced,
as some other body happens thereby to be altered or
destroyed.

Since then those things which chymists produce by the help of the fire are but inanimate bodies; since such fruits of the chymists' skill differ from one another but in so few qualities that we see plainly that by fire, and other agents we can employ, we can easily enough work as great alterations upon matter, as those that are requisite to change one of these chymical productions into another; since the same portion of matter may without being compounded with any extraneous body, or at least element, be made to put on such a variety of formes, and consequently to be (successively) turned into so many differing bodies; and since the matter, cloathed with so many differing formes, was originally but water, and that in its passage through so many transformations, it was never reduced into any of those substances which are reputed to be the principles or elements of mixt bodies, except the violence of the fire, which itself divides not bodies into perfectly simple or elementary substances, but into new compounds; since, I say, these things are so, I see not why we must needs believe that there are any primogeneal and simple bodies, of which, as of pre-existent elements, nature is obliged to compound all others. Nor do I see why we may not conceive that she may produce the bodies accounted mixt out of one another by variously altering and contriving their minute parts, without resolving the matter into any such simple or homogeneous substances as are pretended. Neither, to dispatch, do I see why it should be counted absurd to think, that when a body is resolved by the fire into its supposed simple ingredients, those substances are not true and proper elements, but rather were, as it were, accidentally produced by the fire, which by dissipating a body into minute parts does, if those parts be shut up in close vessels, for the most part necessarily bring them to associate themselves after another manner than before, and so bring them into bodies of such different consistences, as the former texture of the body and concurrent circumstances make such disbanded particles apt to constitute; as experience shews us (and I have both noted it, and proved it already) that as there are some concretes whose parts, when dissipated by fire, are fitted to be put into such schemes of matter as we call oyle, and salt, and spirit; so there are others, such as are especially the greatest part of minerals, whose corpuscles being of another size or figure, or perhaps contrived another way, will not in the fire yeeld bodies of the like consistences, but rather others of differing textures; not to mention, that from gold and some other bodies, we see not that the fire separates any distinct substances at all; nor that even those similar parts of bodies, which the chymists obtain by the fire, are the elements whose names they bear, but compound bodies, upon which, for their resemblance to them in consistence, or some other obvious quality, chymists have been pleased to bestow such appellations.

THE CONCLUSION

THESE last words of Carneades being soon after followed by a noise which seemed to come from the place where the rest of the company was, he took it for a warning, that it was time for him to conclude or break off his discourse; and told his friend; By this time I hope you see, Eleutherius, that if Helmont's experiments be true, it is no absurdity to question whether that doctrine be one, that doth not assert any elements in the sence before explained. But because that, as divers of my arguments suppose the marvellous power of the alkahest in the analyzing of bodies, so the effects ascribed to that power are so unparalleled and stupendous, that though I am not sure but that there may be such an agent, yet little less than ἀυτοψία seems requisite to make a man sure there is. And consequently I leave it to you to judge, how farr those of my arguments that are built upon alkahestical operations are weakned by that liquors being matchless; and shall therefore desire you not to think that I propose this paradox that rejects all elements, as an opinion equally probable with the former part of my discourse. For by that, I hope, you are satisfied, that the arguments, wont to be brought by chymists to prove that all bodies consist of either three principles, or five, are far from being so strong as those that I have employed to prove, that there is not any certain and determinate number of such principles or elements to be met with universally in all mixt bodies. And I suppose I need not tell you, that these anti-chymical paradoxes might have been managed more to their advantage; but that having not confined my curiosity to chymical experiments, I, who am but a young man, and younger chymist, can yet be but slenderly furnished with them, in reference to so great and difficult a task as you imposed upon me: besides that, to tell you the truth, I durst not employ some even of the best experiments I am acquainted with, because I must not yet disclose them; but, however, I think I may presume that what I have hitherto discoursed will induce you to think, that chymists have been much more happy in finding experiments than the causes of them; or in assigning the principles by which they may best be explained. And indeed, when in the writing of Paracelsus I meet with such phantastick and unintelligible discourses as that writer often puzzels and tires his reader with, fathered upon such excellent experiments, as though he seldom clearly teaches, I often find he knew; methinks the chymists, in their searches after truth, are not unlike the navigators of Solomon's Tarshish fleet, who brought home from their long and tedious voyages, not only gold, and silver, and ivory, but apes and peacocks too; for so the writings of several (for I say not, all) of your hermetick philosophers present us, together with divers substantial and noble experiments, theories, which either like peacocks' feathers make a great shew, but are neither solid nor useful; or else like apes, if they have some appearance of being rational, are blemished with some absurdity or other, that when they are attentively considered, make them appear ridiculous.

Carneades having thus finished his discourse against the received doctrines of the elements, Eleutherius judging he should not have time to say much to him before their separation, made some haste to tell him; I confess, Carneades, that you have said more in favour of your paradoxes than I expected. For though divers of the experiments you have mentioned are no secrets, and were not unknown to me, yet besides that you have added many of your own unto them, you have laid them together in such a way, and applyed them to such purposes, and made such deductions from them, as I have not hitherto met with.

But though I be therefore inclined to think, that Philoponus, had he heard you, would scarce have been able in all points to defend the chymical hypothesis against the arguments wherewith you have opposed it; yet methinks that however your objections seem to evince a great part of what they pretend to, yet they evince it not all; and the numerous tryals of those you call the vulgar chymists, may be allowed to prove something too.

Wherefore, if it be granted you that you have made it probable,

First, that the differing substances into which mixt bodies are wont to be resolved by the fire are not of a pure and an elementary nature, especially for this reason, that they yet retain so much of the nature of the concrete that afforded them, as to appear to be yet somewhat compounded, and oftentimes to differ in one concrete from principles of the same denomination in another:

Next, that as to the number of these differing substances, neither is it precisely three, because in most vegetable and animal bodies earth and phlegme are also to be found among their ingredients; nor is there any one determinate number into which the fire (as it is wont to be employed) does precisely and universally resolve all compound bodies whatsoever, as well minerals as others that are reputed perfectly mixt.

Lastly, that there are divers qualities which cannot well be referred to any of these substances, as if they primarily resided in it and belonged to it; and some other qualities, which though they seem to have their chief and most ordinary residence in some one of these principles or elements of mixt bodies, are not yet so deducible from it, but that also some more general principles must be taken in to explicate them.

If, I say, the chymists (continues Eleutherius) be so liberall as to make you these three concessions, I hope you will, on your part, be so civil and equitable as to grant them these three other propositions, namely;

First, that divers mineral bodies, and therefore probably all the rest, may be resolved into a saline, a sulphureous, and a mercurial part; and that almost all vegetable and animal concretes may, if not by the fire alone, yet by a skilfull artist employing the fire as his chief instrument, be divided into five differing substances, salt, spirit, oyle, phlegme and earth; of which the three former by reason of their being so much more operative than the two latter, deserve to be lookt upon as the three active principles, and by way of eminence to be called the three principles of mixt bodies.

Next, that these principles, though they be not perfectly devoid of all mixture, yet may without inconvenience be stiled the elements of compounded bodies, and bear the names of those substances which they most resemble, and which are manifestly predominant in them; and that especially for this reason, that none of these elements is divisible by the fire into four or five differing substances, like the concrete whence it was separated.

Lastly, that divers of the qualities of a mixt body, and especially the medical virtues, do for the most part lodge in some one or other of its principles, and may therefore usefully be sought for in that principle severed from the others.

And in this also (pursues Eleutherius) methinks both you and the chymists may easily agree, that the surest way is to learn by particular experiments, what differing parts particular bodies do consist of, and by what wayes (either actual or potential fire) they may best and most conveniently be separated, as without relying too much upon the fire alone, for the resolving of bodies, so without fruitlessly contending to force them into more elements than nature made them up of, or strip the severed principles so naked, as by making them exquisitely elementary to make them almost useless.

These things (subjoynes Eleu.) I propose, without despairing to see them granted by you; not only because I know that you so much prefer the reputation of candour before that of subtility, that your having once supposed a truth would not hinder you from imbracing it when clearly made out to you; but because, upon the present occasion, it will be no disparagement to you to recede from some of your paradoxes, since the nature and occasion of your past discourse did not oblige you to declare your own opinions, but only to personate an antagonist of the chymists. So that (concludes he, with a smile) you may now by granting what I propose, add the reputation of loving the truth sincerely to that of having been able to oppose it subtilly.

Carneades's haste forbidding him to answer this crafty piece of flattery; Till I shall (saies he) have an opportunity to acquaint you with my own opinions about the controversies I have been discoursing of, you will not I hope, expect I should declare my own sence of the argument I have employed. Wherefore I shall only tell you thus much at present; that though not only an acute naturalist, but even I myself could take plausible exceptions at some of them; yet divers of them too are such as will not perhaps be readily answered, and will reduce my adversaries, at least, to alter and reform their hypothesis. I perceive I need not mind you that the objections I made against the quaternary of elements and ternary of principles needed not to be opposed so much against the doctrines themselves, either of which, especially the latter, may be much more probably maintained than hitherto it seems to have been, by those writers for it I have met with) as against the unaccurateness and the unconcludingness of the analytical experiments vulgarly relyed on to demonstrate them.

And therefore, if either of the two examined opinions, or any other theory of elements, shall upon rational and experimental grounds be clearly made out to me; 'tis obliging, but not irrational, in you to expect, that I shall not be so farr in love with my disquieting doubts, as not to be content to change them for undoubted truths. And (concludes Carneades smiling) it were no great disparagement for a sceptick to confesse to you, that as unsatisfyed as the past discourse may have made you think me with the doctrines of the Peripateticks, and the chymists, about the elements and principles, I can yet so little discover what to acquiesce in, that perchance the enquiries of others have scarce been more unsatisfactory to me, than my own have been to myself.

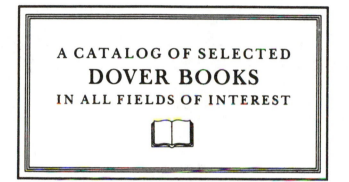

A CATALOG OF SELECTED
DOVER BOOKS
IN ALL FIELDS OF INTEREST

A CATALOG OF SELECTED DOVER
BOOKS IN ALL FIELDS OF INTEREST

CONCERNING THE SPIRITUAL IN ART, Wassily Kandinsky. Pioneering work by father of abstract art. Thoughts on color theory, nature of art. Analysis of earlier masters. 12 illustrations. 80pp. of text. 5⅜ x 8½. 23411-8

ANIMALS: 1,419 Copyright-Free Illustrations of Mammals, Birds, Fish, Insects, etc., Jim Harter (ed.). Clear wood engravings present, in extremely lifelike poses, over 1,000 species of animals. One of the most extensive pictorial sourcebooks of its kind. Captions. Index. 284pp. 9 x 12. 23766-4

CELTIC ART: The Methods of Construction, George Bain. Simple geometric techniques for making Celtic interlacements, spirals, Kells-type initials, animals, humans, etc. Over 500 illustrations. 160pp. 9 x 12. (Available in U.S. only.) 22923-8

AN ATLAS OF ANATOMY FOR ARTISTS, Fritz Schider. Most thorough reference work on art anatomy in the world. Hundreds of illustrations, including selections from works by Vesalius, Leonardo, Goya, Ingres, Michelangelo, others. 593 illustrations. 192pp. 7⅛ x 10¼. 20241-0

CELTIC HAND STROKE-BY-STROKE (Irish Half-Uncial from "The Book of Kells"): An Arthur Baker Calligraphy Manual, Arthur Baker. Complete guide to creating each letter of the alphabet in distinctive Celtic manner. Covers hand position, strokes, pens, inks, paper, more. Illustrated. 48pp. 8¼ x 11. 24336-2

EASY ORIGAMI, John Montroll. Charming collection of 32 projects (hat, cup, pelican, piano, swan, many more) specially designed for the novice origami hobbyist. Clearly illustrated easy-to-follow instructions insure that even beginning papercrafters will achieve successful results. 48pp. 8¼ x 11. 27298-2

THE COMPLETE BOOK OF BIRDHOUSE CONSTRUCTION FOR WOOD-WORKERS, Scott D. Campbell. Detailed instructions, illustrations, tables. Also data on bird habitat and instinct patterns. Bibliography. 3 tables. 63 illustrations in 15 figures. 48pp. 5¼ x 8½. 24407-5

BLOOMINGDALE'S ILLUSTRATED 1886 CATALOG: Fashions, Dry Goods and Housewares, Bloomingdale Brothers. Famed merchants' extremely rare catalog depicting about 1,700 products: clothing, housewares, firearms, dry goods, jewelry, more. Invaluable for dating, identifying vintage items. Also, copyright-free graphics for artists, designers. Co-published with Henry Ford Museum & Greenfield Village. 160pp. 8¼ x 11. 25780-0

HISTORIC COSTUME IN PICTURES, Braun & Schneider. Over 1,450 costumed figures in clearly detailed engravings–from dawn of civilization to end of 19th century. Captions. Many folk costumes. 256pp. 8⅜ x 11¾. 23150-X

STICKLEY CRAFTSMAN FURNITURE CATALOGS, Gustav Stickley and L. & J. G. Stickley. Beautiful, functional furniture in two authentic catalogs from 1910. 594 illustrations, including 277 photos, show settles, rockers, armchairs, reclining chairs, bookcases, desks, tables. 183pp. 6½ x 9¼. 23838-5

AMERICAN LOCOMOTIVES IN HISTORIC PHOTOGRAPHS: 1858 to 1949, Ron Ziel (ed.). A rare collection of 126 meticulously detailed official photographs, called "builder portraits," of American locomotives that majestically chronicle the rise of steam locomotive power in America. Introduction. Detailed captions. xi+ 129pp. 9 x 12. 27393-8

AMERICA'S LIGHTHOUSES: An Illustrated History, Francis Ross Holland, Jr. Delightfully written, profusely illustrated fact-filled survey of over 200 American lighthouses since 1716. History, anecdotes, technological advances, more. 240pp. 8 x 10¾. 25576-X

TOWARDS A NEW ARCHITECTURE, Le Corbusier. Pioneering manifesto by founder of "International School." Technical and aesthetic theories, views of industry, economics, relation of form to function, "mass-production split" and much more. Profusely illustrated. 320pp. 6⅛ x 9¼. (Available in U.S. only.) 25023-7

HOW THE OTHER HALF LIVES, Jacob Riis. Famous journalistic record, exposing poverty and degradation of New York slums around 1900, by major social reformer. 100 striking and influential photographs. 233pp. 10 x 7⅞. 22012-5

FRUIT KEY AND TWIG KEY TO TREES AND SHRUBS, William M. Harlow. One of the handiest and most widely used identification aids. Fruit key covers 120 deciduous and evergreen species; twig key 160 deciduous species. Easily used. Over 300 photographs. 126pp. 5⅜ x 8½. 20511-8

COMMON BIRD SONGS, Dr. Donald J. Borror. Songs of 60 most common U.S. birds: robins, sparrows, cardinals, bluejays, finches, more–arranged in order of increasing complexity. Up to 9 variations of songs of each species.
Cassette and manual 99911-4

ORCHIDS AS HOUSE PLANTS, Rebecca Tyson Northen. Grow cattleyas and many other kinds of orchids–in a window, in a case, or under artificial light. 63 illustrations. 148pp. 5⅜ x 8½. 23261-1

MONSTER MAZES, Dave Phillips. Masterful mazes at four levels of difficulty. Avoid deadly perils and evil creatures to find magical treasures. Solutions for all 32 exciting illustrated puzzles. 48pp. 8¼ x 11. 26005-4

MOZART'S DON GIOVANNI (DOVER OPERA LIBRETTO SERIES), Wolfgang Amadeus Mozart. Introduced and translated by Ellen H. Bleiler. Standard Italian libretto, with complete English translation. Convenient and thoroughly portable–an ideal companion for reading along with a recording or the performance itself. Introduction. List of characters. Plot summary. 121pp. 5¼ x 8½. 24944-1

TECHNICAL MANUAL AND DICTIONARY OF CLASSICAL BALLET, Gail Grant. Defines, explains, comments on steps, movements, poses and concepts. 15-page pictorial section. Basic book for student, viewer. 127pp. 5⅜ x 8½. 21843-0

THE CLARINET AND CLARINET PLAYING, David Pino. Lively, comprehensive work features suggestions about technique, musicianship, and musical interpretation, as well as guidelines for teaching, making your own reeds, and preparing for public performance. Includes an intriguing look at clarinet history. "A godsend," *The Clarinet,* Journal of the International Clarinet Society. Appendixes. 7 illus. 320pp. 5⅜ x 8½. 40270-3

HOLLYWOOD GLAMOR PORTRAITS, John Kobal (ed.). 145 photos from 1926-49. Harlow, Gable, Bogart, Bacall; 94 stars in all. Full background on photographers, technical aspects. 160pp. 8⅜ x 11¼. 23352-9

THE ANNOTATED CASEY AT THE BAT: A Collection of Ballads about the Mighty Casey/Third, Revised Edition, Martin Gardner (ed.). Amusing sequels and parodies of one of America's best-loved poems: Casey's Revenge, Why Casey Whiffed, Casey's Sister at the Bat, others. 256pp. 5⅜ x 8½. 28598-7

THE RAVEN AND OTHER FAVORITE POEMS, Edgar Allan Poe. Over 40 of the author's most memorable poems: "The Bells," "Ulalume," "Israfel," "To Helen," "The Conqueror Worm," "Eldorado," "Annabel Lee," many more. Alphabetic lists of titles and first lines. 64pp. 5³⁄₁₆ x 8¼. 26685-0

PERSONAL MEMOIRS OF U. S. GRANT, Ulysses Simpson Grant. Intelligent, deeply moving firsthand account of Civil War campaigns, considered by many the finest military memoirs ever written. Includes letters, historic photographs, maps and more. 528pp. 6⅛ x 9¼. 28587-1

ANCIENT EGYPTIAN MATERIALS AND INDUSTRIES, A. Lucas and J. Harris. Fascinating, comprehensive, thoroughly documented text describes this ancient civilization's vast resources and the processes that incorporated them in daily life, including the use of animal products, building materials, cosmetics, perfumes and incense, fibers, glazed ware, glass and its manufacture, materials used in the mummification process, and much more. 544pp. 6⅛ x 9¼. (Available in U.S. only.) 40446-3

RUSSIAN STORIES/RUSSKIE RASSKAZY: A Dual-Language Book, edited by Gleb Struve. Twelve tales by such masters as Chekhov, Tolstoy, Dostoevsky, Pushkin, others. Excellent word-for-word English translations on facing pages, plus teaching and study aids, Russian/English vocabulary, biographical/critical introductions, more. 416pp. 5⅜ x 8½. 26244-8

PHILADELPHIA THEN AND NOW: 60 Sites Photographed in the Past and Present, Kenneth Finkel and Susan Oyama. Rare photographs of City Hall, Logan Square, Independence Hall, Betsy Ross House, other landmarks juxtaposed with contemporary views. Captures changing face of historic city. Introduction. Captions. 128pp. 8¼ x 11. 25790-8

AIA ARCHITECTURAL GUIDE TO NASSAU AND SUFFOLK COUNTIES, LONG ISLAND, The American Institute of Architects, Long Island Chapter, and the Society for the Preservation of Long Island Antiquities. Comprehensive, well-researched and generously illustrated volume brings to life over three centuries of Long Island's great architectural heritage. More than 240 photographs with authoritative, extensively detailed captions. 176pp. 8¼ x 11. 26946-9

NORTH AMERICAN INDIAN LIFE: Customs and Traditions of 23 Tribes, Elsie Clews Parsons (ed.). 27 fictionalized essays by noted anthropologists examine religion, customs, government, additional facets of life among the Winnebago, Crow, Zuni, Eskimo, other tribes. 480pp. 6⅛ x 9¼. 27377-6

CATALOG OF DOVER BOOKS

FRANK LLOYD WRIGHT'S DANA HOUSE, Donald Hoffmann. Pictorial essay of residential masterpiece with over 160 interior and exterior photos, plans, elevations, sketches and studies. 128pp. 9¼ x 10¾. 29120-0

THE MALE AND FEMALE FIGURE IN MOTION: 60 Classic Photographic Sequences, Eadweard Muybridge. 60 true-action photographs of men and women walking, running, climbing, bending, turning, etc., reproduced from rare 19th-century masterpiece. vi + 121pp. 9 x 12. 24745-7

1001 QUESTIONS ANSWERED ABOUT THE SEASHORE, N. J. Berrill and Jacquelyn Berrill. Queries answered about dolphins, sea snails, sponges, starfish, fishes, shore birds, many others. Covers appearance, breeding, growth, feeding, much more. 305pp. 5¼ x 8¼. 23366-9

ATTRACTING BIRDS TO YOUR YARD, William J. Weber. Easy-to-follow guide offers advice on how to attract the greatest diversity of birds: birdhouses, feeders, water and waterers, much more. 96pp. 5³⁄₁₆ x 8¼. 28927-3

MEDICINAL AND OTHER USES OF NORTH AMERICAN PLANTS: A Historical Survey with Special Reference to the Eastern Indian Tribes, Charlotte Erichsen-Brown. Chronological historical citations document 500 years of usage of plants, trees, shrubs native to eastern Canada, northeastern U.S. Also complete identifying information. 343 illustrations. 544pp. 6½ x 9¼. 25951-X

STORYBOOK MAZES, Dave Phillips. 23 stories and mazes on two-page spreads: Wizard of Oz, Treasure Island, Robin Hood, etc. Solutions. 64pp. 8¼ x 11. 23628-5

AMERICAN NEGRO SONGS: 230 Folk Songs and Spirituals, Religious and Secular, John W. Work. This authoritative study traces the African influences of songs sung and played by black Americans at work, in church, and as entertainment. The author discusses the lyric significance of such songs as "Swing Low, Sweet Chariot," "John Henry," and others and offers the words and music for 230 songs. Bibliography. Index of Song Titles. 272pp. 6½ x 9¼. 40271-1

MOVIE-STAR PORTRAITS OF THE FORTIES, John Kobal (ed.). 163 glamor, studio photos of 106 stars of the 1940s: Rita Hayworth, Ava Gardner, Marlon Brando, Clark Gable, many more. 176pp. 8⅜ x 11¼. 23546-7

BENCHLEY LOST AND FOUND, Robert Benchley. Finest humor from early 30s, about pet peeves, child psychologists, post office and others. Mostly unavailable elsewhere. 73 illustrations by Peter Arno and others. 183pp. 5⅜ x 8½. 22410-4

YEKL and THE IMPORTED BRIDEGROOM AND OTHER STORIES OF YIDDISH NEW YORK, Abraham Cahan. Film Hester Street based on *Yekl* (1896). Novel, other stories among first about Jewish immigrants on N.Y.'s East Side. 240pp. 5⅜ x 8½. 22427-9

SELECTED POEMS, Walt Whitman. Generous sampling from *Leaves of Grass*. Twenty-four poems include "I Hear America Singing," "Song of the Open Road," "I Sing the Body Electric," "When Lilacs Last in the Dooryard Bloom'd," "O Captain! My Captain!"–all reprinted from an authoritative edition. Lists of titles and first lines. 128pp. 5³⁄₁₆ x 8¼. 26878-0

THE BEST TALES OF HOFFMANN, E. T. A. Hoffmann. 10 of Hoffmann's most important stories: "Nutcracker and the King of Mice," "The Golden Flowerpot," etc. 458pp. 5⅜ x 8½. 21793-0

FROM FETISH TO GOD IN ANCIENT EGYPT, E. A. Wallis Budge. Rich detailed survey of Egyptian conception of "God" and gods, magic, cult of animals, Osiris, more. Also, superb English translations of hymns and legends. 240 illustrations. 545pp. 5⅜ x 8½. 25803-3

FRENCH STORIES/CONTES FRANÇAIS: A Dual-Language Book, Wallace Fowlie. Ten stories by French masters, Voltaire to Camus: "Micromegas" by Voltaire; "The Atheist's Mass" by Balzac; "Minuet" by de Maupassant; "The Guest" by Camus, six more. Excellent English translations on facing pages. Also French-English vocabulary list, exercises, more. 352pp. 5⅜ x 8½. 26443-2

CHICAGO AT THE TURN OF THE CENTURY IN PHOTOGRAPHS: 122 Historic Views from the Collections of the Chicago Historical Society, Larry A. Viskochil. Rare large-format prints offer detailed views of City Hall, State Street, the Loop, Hull House, Union Station, many other landmarks, circa 1904-1913. Introduction. Captions. Maps. 144pp. 9⅜ x 12¼. 24656-6

OLD BROOKLYN IN EARLY PHOTOGRAPHS, 1865-1929, William Lee Younger. Luna Park, Gravesend race track, construction of Grand Army Plaza, moving of Hotel Brighton, etc. 157 previously unpublished photographs. 165pp. 8⅞ x 11¾. 23587-4

THE MYTHS OF THE NORTH AMERICAN INDIANS, Lewis Spence. Rich anthology of the myths and legends of the Algonquins, Iroquois, Pawnees and Sioux, prefaced by an extensive historical and ethnological commentary. 36 illustrations. 480pp. 5⅜ x 8½. 25967-6

AN ENCYCLOPEDIA OF BATTLES: Accounts of Over 1,560 Battles from 1479 B.C. to the Present, David Eggenberger. Essential details of every major battle in recorded history from the first battle of Megiddo in 1479 B.C. to Grenada in 1984. List of Battle Maps. New Appendix covering the years 1967-1984. Index. 99 illustrations. 544pp. 6½ x 9¼. 24913-1

SAILING ALONE AROUND THE WORLD, Captain Joshua Slocum. First man to sail around the world, alone, in small boat. One of great feats of seamanship told in delightful manner. 67 illustrations. 294pp. 5⅜ x 8½. 20326-3

ANARCHISM AND OTHER ESSAYS, Emma Goldman. Powerful, penetrating, prophetic essays on direct action, role of minorities, prison reform, puritan hypocrisy, violence, etc. 271pp. 5⅜ x 8½. 22484-8

MYTHS OF THE HINDUS AND BUDDHISTS, Ananda K. Coomaraswamy and Sister Nivedita. Great stories of the epics; deeds of Krishna, Shiva, taken from puranas, Vedas, folk tales; etc. 32 illustrations. 400pp. 5⅜ x 8½. 21759-0

THE TRAUMA OF BIRTH, Otto Rank. Rank's controversial thesis that anxiety neurosis is caused by profound psychological trauma which occurs at birth. 256pp. 5⅜ x 8½. 27974-X

A THEOLOGICO-POLITICAL TREATISE, Benedict Spinoza. Also contains unfinished Political Treatise. Great classic on religious liberty, theory of government on common consent. R. Elwes translation. Total of 421pp. 5⅜ x 8½. 20249-6

MY BONDAGE AND MY FREEDOM, Frederick Douglass. Born a slave, Douglass became outspoken force in antislavery movement. The best of Douglass' autobiographies. Graphic description of slave life. 464pp. 5⅜ x 8½. 22457-0

FOLLOWING THE EQUATOR: A Journey Around the World, Mark Twain. Fascinating humorous account of 1897 voyage to Hawaii, Australia, India, New Zealand, etc. Ironic, bemused reports on peoples, customs, climate, flora and fauna, politics, much more. 197 illustrations. 720pp. 5⅜ x 8½. 26113-1

THE PEOPLE CALLED SHAKERS, Edward D. Andrews. Definitive study of Shakers: origins, beliefs, practices, dances, social organization, furniture and crafts, etc. 33 illustrations. 351pp. 5⅜ x 8½. 21081-2

THE MYTHS OF GREECE AND ROME, H. A. Guerber. A classic of mythology, generously illustrated, long prized for its simple, graphic, accurate retelling of the principal myths of Greece and Rome, and for its commentary on their origins and significance. With 64 illustrations by Michelangelo, Raphael, Titian, Rubens, Canova, Bernini and others. 480pp. 5⅜ x 8½. 27584-1

PSYCHOLOGY OF MUSIC, Carl E. Seashore. Classic work discusses music as a medium from psychological viewpoint. Clear treatment of physical acoustics, auditory apparatus, sound perception, development of musical skills, nature of musical feeling, host of other topics. 88 figures. 408pp. 5⅜ x 8½. 21851-1

THE PHILOSOPHY OF HISTORY, Georg W. Hegel. Great classic of Western thought develops concept that history is not chance but rational process, the evolution of freedom. 457pp. 5⅜ x 8½. 20112-0

THE BOOK OF TEA, Kakuzo Okakura. Minor classic of the Orient: entertaining, charming explanation, interpretation of traditional Japanese culture in terms of tea ceremony. 94pp. 5⅜ x 8½. 20070-1

LIFE IN ANCIENT EGYPT, Adolf Erman. Fullest, most thorough, detailed older account with much not in more recent books, domestic life, religion, magic, medicine, commerce, much more. Many illustrations reproduce tomb paintings, carvings, hieroglyphs, etc. 597pp. 5⅜ x 8½. 22632-8

SUNDIALS, Their Theory and Construction, Albert Waugh. Far and away the best, most thorough coverage of ideas, mathematics concerned, types, construction, adjusting anywhere. Simple, nontechnical treatment allows even children to build several of these dials. Over 100 illustrations. 230pp. 5⅜ x 8½. 22947-5

THEORETICAL HYDRODYNAMICS, L. M. Milne-Thomson. Classic exposition of the mathematical theory of fluid motion, applicable to both hydrodynamics and aerodynamics. Over 600 exercises. 768pp. 6⅛ x 9¼. 68970-0

SONGS OF EXPERIENCE: Facsimile Reproduction with 26 Plates in Full Color, William Blake. 26 full-color plates from a rare 1826 edition. Includes "The Tyger," "London," "Holy Thursday," and other poems. Printed text of poems. 48pp. 5¼ x 7. 24636-1

OLD-TIME VIGNETTES IN FULL COLOR, Carol Belanger Grafton (ed.). Over 390 charming, often sentimental illustrations, selected from archives of Victorian graphics—pretty women posing, children playing, food, flowers, kittens and puppies, smiling cherubs, birds and butterflies, much more. All copyright-free. 48pp. 9¼ x 12¼. 27269-9

PERSPECTIVE FOR ARTISTS, Rex Vicat Cole. Depth, perspective of sky and sea, shadows, much more, not usually covered. 391 diagrams, 81 reproductions of drawings and paintings. 279pp. 5⅜ x 8½. 22487-2

DRAWING THE LIVING FIGURE, Joseph Sheppard. Innovative approach to artistic anatomy focuses on specifics of surface anatomy, rather than muscles and bones. Over 170 drawings of live models in front, back and side views, and in widely varying poses. Accompanying diagrams. 177 illustrations. Introduction. Index. 144pp. 8⅜ x11¼. 26723-7

GOTHIC AND OLD ENGLISH ALPHABETS: 100 Complete Fonts, Dan X. Solo. Add power, elegance to posters, signs, other graphics with 100 stunning copyright-free alphabets: Blackstone, Dolbey, Germania, 97 more–including many lower-case, numerals, punctuation marks. 104pp. 8⅛ x 11. 24695-7

HOW TO DO BEADWORK, Mary White. Fundamental book on craft from simple projects to five-bead chains and woven works. 106 illustrations. 142pp. 5⅜ x 8. 20697-1

THE BOOK OF WOOD CARVING, Charles Marshall Sayers. Finest book for beginners discusses fundamentals and offers 34 designs. "Absolutely first rate . . . well thought out and well executed."–E. J. Tangerman. 118pp. 7¾ x 10⅝. 23654-4

ILLUSTRATED CATALOG OF CIVIL WAR MILITARY GOODS: Union Army Weapons, Insignia, Uniform Accessories, and Other Equipment, Schuyler, Hartley, and Graham. Rare, profusely illustrated 1846 catalog includes Union Army uniform and dress regulations, arms and ammunition, coats, insignia, flags, swords, rifles, etc. 226 illustrations. 160pp. 9 x 12. 24939-5

WOMEN'S FASHIONS OF THE EARLY 1900s: An Unabridged Republication of "New York Fashions, 1909," National Cloak & Suit Co. Rare catalog of mail-order fashions documents women's and children's clothing styles shortly after the turn of the century. Captions offer full descriptions, prices. Invaluable resource for fashion, costume historians. Approximately 725 illustrations. 128pp. 8⅜ x 11¼. 27276-1

THE 1912 AND 1915 GUSTAV STICKLEY FURNITURE CATALOGS, Gustav Stickley. With over 200 detailed illustrations and descriptions, these two catalogs are essential reading and reference materials and identification guides for Stickley furniture. Captions cite materials, dimensions and prices. 112pp. 6½ x 9¼. 26676-1

EARLY AMERICAN LOCOMOTIVES, John H. White, Jr. Finest locomotive engravings from early 19th century: historical (1804–74), main-line (after 1870), special, foreign, etc. 147 plates. 142pp. 11⅜ x 8¼. 22772-3

THE TALL SHIPS OF TODAY IN PHOTOGRAPHS, Frank O. Braynard. Lavishly illustrated tribute to nearly 100 majestic contemporary sailing vessels: Amerigo Vespucci, Clearwater, Constitution, Eagle, Mayflower, Sea Cloud, Victory, many more. Authoritative captions provide statistics, background on each ship. 190 black-and-white photographs and illustrations. Introduction. 128pp. 8⅞ x 11¾. 27163-3

LITTLE BOOK OF EARLY AMERICAN CRAFTS AND TRADES, Peter Stockham (ed.). 1807 children's book explains crafts and trades: baker, hatter, cooper, potter, and many others. 23 copperplate illustrations. 140pp. 4⅝ x 6.   23336-7

VICTORIAN FASHIONS AND COSTUMES FROM HARPER'S BAZAR, 1867–1898, Stella Blum (ed.). Day costumes, evening wear, sports clothes, shoes, hats, other accessories in over 1,000 detailed engravings. 320pp. 9⅜ x 12¼.   22990-4

GUSTAV STICKLEY, THE CRAFTSMAN, Mary Ann Smith. Superb study surveys broad scope of Stickley's achievement, especially in architecture. Design philosophy, rise and fall of the Craftsman empire, descriptions and floor plans for many Craftsman houses, more. 86 black-and-white halftones. 31 line illustrations. Introduction 208pp. 6½ x 9¼.   27210-9

THE LONG ISLAND RAIL ROAD IN EARLY PHOTOGRAPHS, Ron Ziel. Over 220 rare photos, informative text document origin (1844) and development of rail service on Long Island. Vintage views of early trains, locomotives, stations, passengers, crews, much more. Captions. 8⅞ x 11¾.   26301-0

VOYAGE OF THE LIBERDADE, Joshua Slocum. Great 19th-century mariner's thrilling, first-hand account of the wreck of his ship off South America, the 35-foot boat he built from the wreckage, and its remarkable voyage home. 128pp. 5⅜ x 8½.   40022-0

TEN BOOKS ON ARCHITECTURE, Vitruvius. The most important book ever written on architecture. Early Roman aesthetics, technology, classical orders, site selection, all other aspects. Morgan translation. 331pp. 5⅜ x 8½.   20645-9

THE HUMAN FIGURE IN MOTION, Eadweard Muybridge. More than 4,500 stopped-action photos, in action series, showing undraped men, women, children jumping, lying down, throwing, sitting, wrestling, carrying, etc. 390pp. 7⅞ x 10⅝.   20204-6 Clothbd.

TREES OF THE EASTERN AND CENTRAL UNITED STATES AND CANADA, William M. Harlow. Best one-volume guide to 140 trees. Full descriptions, woodlore, range, etc. Over 600 illustrations. Handy size. 288pp. 4½ x 6⅜.   20395-6

SONGS OF WESTERN BIRDS, Dr. Donald J. Borror. Complete song and call repertoire of 60 western species, including flycatchers, juncoes, cactus wrens, many more—includes fully illustrated booklet.   Cassette and manual 99913-0

GROWING AND USING HERBS AND SPICES, Milo Miloradovich. Versatile handbook provides all the information needed for cultivation and use of all the herbs and spices available in North America. 4 illustrations. Index. Glossary. 236pp. 5⅜ x 8½.   25058-X

BIG BOOK OF MAZES AND LABYRINTHS, Walter Shepherd. 50 mazes and labyrinths in all—classical, solid, ripple, and more—in one great volume. Perfect inexpensive puzzler for clever youngsters. Full solutions. 112pp. 8⅛ x 11.   22951-3

CATALOG OF DOVER BOOKS

PIANO TUNING, J. Cree Fischer. Clearest, best book for beginner, amateur. Simple repairs, raising dropped notes, tuning by easy method of flattened fifths. No previous skills needed. 4 illustrations. 201pp. 5⅜ x 8½.   23267-0

HINTS TO SINGERS, Lillian Nordica. Selecting the right teacher, developing confidence, overcoming stage fright, and many other important skills receive thoughtful discussion in this indispensible guide, written by a world-famous diva of four decades' experience. 96pp. 5⅜ x 8½.   40094-8

THE COMPLETE NONSENSE OF EDWARD LEAR, Edward Lear. All nonsense limericks, zany alphabets, Owl and Pussycat, songs, nonsense botany, etc., illustrated by Lear. Total of 320pp. 5⅜ x 8½. (Available in U.S. only.)   20167-8

VICTORIAN PARLOUR POETRY: An Annotated Anthology, Michael R. Turner. 117 gems by Longfellow, Tennyson, Browning, many lesser-known poets. "The Village Blacksmith," "Curfew Must Not Ring Tonight," "Only a Baby Small," dozens more, often difficult to find elsewhere. Index of poets, titles, first lines. xxiii + 325pp. 5⅜ x 8¼.   27044-0

DUBLINERS, James Joyce. Fifteen stories offer vivid, tightly focused observations of the lives of Dublin's poorer classes. At least one, "The Dead," is considered a masterpiece. Reprinted complete and unabridged from standard edition. 160pp. 5³⁄₁₆ x 8¼.   26870-5

GREAT WEIRD TALES: 14 Stories by Lovecraft, Blackwood, Machen and Others, S. T. Joshi (ed.). 14 spellbinding tales, including "The Sin Eater," by Fiona McLeod, "The Eye Above the Mantel," by Frank Belknap Long, as well as renowned works by R. H. Barlow, Lord Dunsany, Arthur Machen, W. C. Morrow and eight other masters of the genre. 256pp. 5⅜ x 8½. (Available in U.S. only.)   40436-6

THE BOOK OF THE SACRED MAGIC OF ABRAMELIN THE MAGE, translated by S. MacGregor Mathers. Medieval manuscript of ceremonial magic. Basic document in Aleister Crowley, Golden Dawn groups. 268pp. 5⅜ x 8½.   23211-5

NEW RUSSIAN-ENGLISH AND ENGLISH-RUSSIAN DICTIONARY, M. A. O'Brien. This is a remarkably handy Russian dictionary, containing a surprising amount of information, including over 70,000 entries. 366pp. 4½ x 6⅛.   20208-9

HISTORIC HOMES OF THE AMERICAN PRESIDENTS, Second, Revised Edition, Irvin Haas. A traveler's guide to American Presidential homes, most open to the public, depicting and describing homes occupied by every American President from George Washington to George Bush. With visiting hours, admission charges, travel routes. 175 photographs. Index. 160pp. 8¼ x 11.   26751-2

NEW YORK IN THE FORTIES, Andreas Feininger. 162 brilliant photographs by the well-known photographer, formerly with *Life* magazine. Commuters, shoppers, Times Square at night, much else from city at its peak. Captions by John von Hartz. 181pp. 9¼ x 10¾.   23585-8

INDIAN SIGN LANGUAGE, William Tomkins. Over 525 signs developed by Sioux and other tribes. Written instructions and diagrams. Also 290 pictographs. 111pp. 6⅛ x 9¼.   22029-X

ANATOMY: A Complete Guide for Artists, Joseph Sheppard. A master of figure drawing shows artists how to render human anatomy convincingly. Over 460 illustrations. 224pp. 8⅜ x 11¼. 27279-6

MEDIEVAL CALLIGRAPHY: Its History and Technique, Marc Drogin. Spirited history, comprehensive instruction manual covers 13 styles (ca. 4th century through 15th). Excellent photographs; directions for duplicating medieval techniques with modern tools. 224pp. 8⅜ x 11¼. 26142-5

DRIED FLOWERS: How to Prepare Them, Sarah Whitlock and Martha Rankin. Complete instructions on how to use silica gel, meal and borax, perlite aggregate, sand and borax, glycerine and water to create attractive permanent flower arrangements. 12 illustrations. 32pp. 5⅜ x 8½. 21802-3

EASY-TO-MAKE BIRD FEEDERS FOR WOODWORKERS, Scott D. Campbell. Detailed, simple-to-use guide for designing, constructing, caring for and using feeders. Text, illustrations for 12 classic and contemporary designs. 96pp. 5⅜ x 8½. 25847-5

SCOTTISH WONDER TALES FROM MYTH AND LEGEND, Donald A. Mackenzie. 16 lively tales tell of giants rumbling down mountainsides, of a magic wand that turns stone pillars into warriors, of gods and goddesses, evil hags, powerful forces and more. 240pp. 5⅜ x 8½. 29677-6

THE HISTORY OF UNDERCLOTHES, C. Willett Cunnington and Phyllis Cunnington. Fascinating, well-documented survey covering six centuries of English undergarments, enhanced with over 100 illustrations: 12th-century laced-up bodice, footed long drawers (1795), 19th-century bustles, 19th-century corsets for men, Victorian "bust improvers," much more. 272pp. 5⅝ x 8¼. 27124-2

ARTS AND CRAFTS FURNITURE: The Complete Brooks Catalog of 1912, Brooks Manufacturing Co. Photos and detailed descriptions of more than 150 now very collectible furniture designs from the Arts and Crafts movement depict davenports, settees, buffets, desks, tables, chairs, bedsteads, dressers and more, all built of solid, quarter-sawed oak. Invaluable for students and enthusiasts of antiques, Americana and the decorative arts. 80pp. 6½ x 9¼. 27471-3

WILBUR AND ORVILLE: A Biography of the Wright Brothers, Fred Howard. Definitive, crisply written study tells the full story of the brothers' lives and work. A vividly written biography, unparalleled in scope and color, that also captures the spirit of an extraordinary era. 560pp. 6⅛ x 9¼. 40297-5

THE ARTS OF THE SAILOR: Knotting, Splicing and Ropework, Hervey Garrett Smith. Indispensable shipboard reference covers tools, basic knots and useful hitches; handsewing and canvas work, more. Over 100 illustrations. Delightful reading for sea lovers. 256pp. 5⅜ x 8½. 26440-8

FRANK LLOYD WRIGHT'S FALLINGWATER: The House and Its History, Second, Revised Edition, Donald Hoffmann. A total revision–both in text and illustrations–of the standard document on Fallingwater, the boldest, most personal architectural statement of Wright's mature years, updated with valuable new material from the recently opened Frank Lloyd Wright Archives. "Fascinating"–*The New York Times*. 116 illustrations. 128pp. 9¼ x 10¾. 27430-6

PHOTOGRAPHIC SKETCHBOOK OF THE CIVIL WAR, Alexander Gardner. 100 photos taken on field during the Civil War. Famous shots of Manassas Harper's Ferry, Lincoln, Richmond, slave pens, etc. 244pp. 10⅝ x 8¼. 22731-6

FIVE ACRES AND INDEPENDENCE, Maurice G. Kains. Great back-to-the-land classic explains basics of self-sufficient farming. The one book to get. 95 illustrations. 397pp. 5⅜ x 8½. 20974-1

SONGS OF EASTERN BIRDS, Dr. Donald J. Borror. Songs and calls of 60 species most common to eastern U.S.: warblers, woodpeckers, flycatchers, thrushes, larks, many more in high-quality recording. Cassette and manual 99912-2

A MODERN HERBAL, Margaret Grieve. Much the fullest, most exact, most useful compilation of herbal material. Gigantic alphabetical encyclopedia, from aconite to zedoary, gives botanical information, medical properties, folklore, economic uses, much else. Indispensable to serious reader. 161 illustrations. 888pp. 6½ x 9¼. 2-vol. set. (Available in U.S. only.) Vol. I: 22798-7
Vol. II: 22799-5

HIDDEN TREASURE MAZE BOOK, Dave Phillips. Solve 34 challenging mazes accompanied by heroic tales of adventure. Evil dragons, people-eating plants, bloodthirsty giants, many more dangerous adversaries lurk at every twist and turn. 34 mazes, stories, solutions. 48pp. 8¼ x 11. 24566-7

LETTERS OF W. A. MOZART, Wolfgang A. Mozart. Remarkable letters show bawdy wit, humor, imagination, musical insights, contemporary musical world; includes some letters from Leopold Mozart. 276pp. 5⅜ x 8½. 22859-2

BASIC PRINCIPLES OF CLASSICAL BALLET, Agrippina Vaganova. Great Russian theoretician, teacher explains methods for teaching classical ballet. 118 illustrations. 175pp. 5⅜ x 8½. 22036-2

THE JUMPING FROG, Mark Twain. Revenge edition. The original story of The Celebrated Jumping Frog of Calaveras County, a hapless French translation, and Twain's hilarious "retranslation" from the French. 12 illustrations. 66pp. 5⅜ x 8½. 22686-7

BEST REMEMBERED POEMS, Martin Gardner (ed.). The 126 poems in this superb collection of 19th- and 20th-century British and American verse range from Shelley's "To a Skylark" to the impassioned "Renascence" of Edna St. Vincent Millay and to Edward Lear's whimsical "The Owl and the Pussycat." 224pp. 5⅜ x 8½. 27165-X

COMPLETE SONNETS, William Shakespeare. Over 150 exquisite poems deal with love, friendship, the tyranny of time, beauty's evanescence, death and other themes in language of remarkable power, precision and beauty. Glossary of archaic terms. 80pp. 5³⁄₁₆ x 8¼. 26686-9

THE BATTLES THAT CHANGED HISTORY, Fletcher Pratt. Eminent historian profiles 16 crucial conflicts, ancient to modern, that changed the course of civilization. 352pp. 5⅜ x 8½. 41129-X

THE WIT AND HUMOR OF OSCAR WILDE, Alvin Redman (ed.). More than 1,000 ripostes, paradoxes, wisecracks: Work is the curse of the drinking classes; I can resist everything except temptation; etc. 258pp. 5⅜ x 8½. 20602-5

SHAKESPEARE LEXICON AND QUOTATION DICTIONARY, Alexander Schmidt. Full definitions, locations, shades of meaning in every word in plays and poems. More than 50,000 exact quotations. 1,485pp. 6½ x 9¼. 2-vol. set.
Vol. 1: 22726-X
Vol. 2: 22727-8

SELECTED POEMS, Emily Dickinson. Over 100 best-known, best-loved poems by one of America's foremost poets, reprinted from authoritative early editions. No comparable edition at this price. Index of first lines. 64pp. 5³⁄₁₆ x 8¼. 26466-1

THE INSIDIOUS DR. FU-MANCHU, Sax Rohmer. The first of the popular mystery series introduces a pair of English detectives to their archnemesis, the diabolical Dr. Fu-Manchu. Flavorful atmosphere, fast-paced action, and colorful characters enliven this classic of the genre. 208pp. 5³⁄₁₆ x 8¼. 29898-1

THE MALLEUS MALEFICARUM OF KRAMER AND SPRENGER, translated by Montague Summers. Full text of most important witchhunter's "bible," used by both Catholics and Protestants. 278pp. 6⅝ x 10. 22802-9

SPANISH STORIES/CUENTOS ESPAÑOLES: A Dual-Language Book, Angel Flores (ed.). Unique format offers 13 great stories in Spanish by Cervantes, Borges, others. Faithful English translations on facing pages. 352pp. 5⅜ x 8½. 25399-6

GARDEN CITY, LONG ISLAND, IN EARLY PHOTOGRAPHS, 1869–1919, Mildred H. Smith. Handsome treasury of 118 vintage pictures, accompanied by carefully researched captions, document the Garden City Hotel fire (1899), the Vanderbilt Cup Race (1908), the first airmail flight departing from the Nassau Boulevard Aerodrome (1911), and much more. 96pp. 8⅞ x 11¾. 40669-5

OLD QUEENS, N.Y., IN EARLY PHOTOGRAPHS, Vincent F. Seyfried and William Asadorian. Over 160 rare photographs of Maspeth, Jamaica, Jackson Heights, and other areas. Vintage views of DeWitt Clinton mansion, 1939 World's Fair and more. Captions. 192pp. 8⅞ x 11. 26358-4

CAPTURED BY THE INDIANS: 15 Firsthand Accounts, 1750-1870, Frederick Drimmer. Astounding true historical accounts of grisly torture, bloody conflicts, relentless pursuits, miraculous escapes and more, by people who lived to tell the tale. 384pp. 5⅜ x 8½. 24901-8

THE WORLD'S GREAT SPEECHES (Fourth Enlarged Edition), Lewis Copeland, Lawrence W. Lamm, and Stephen J. McKenna. Nearly 300 speeches provide public speakers with a wealth of updated quotes and inspiration–from Pericles' funeral oration and William Jennings Bryan's "Cross of Gold Speech" to Malcolm X's powerful words on the Black Revolution and Earl of Spenser's tribute to his sister, Diana, Princess of Wales. 944pp. 5⅜ x 8⅜. 40903-1

THE BOOK OF THE SWORD, Sir Richard F. Burton. Great Victorian scholar/adventurer's eloquent, erudite history of the "queen of weapons"–from prehistory to early Roman Empire. Evolution and development of early swords, variations (sabre, broadsword, cutlass, scimitar, etc.), much more. 336pp. 6⅛ x 9¼. 25434-8

AUTOBIOGRAPHY: The Story of My Experiments with Truth, Mohandas K. Gandhi. Boyhood, legal studies, purification, the growth of the Satyagraha (nonviolent protest) movement. Critical, inspiring work of the man responsible for the freedom of India. 480pp. 5⅜ x 8½. (Available in U.S. only.)   24593-4

CELTIC MYTHS AND LEGENDS, T. W. Rolleston. Masterful retelling of Irish and Welsh stories and tales. Cuchulain, King Arthur, Deirdre, the Grail, many more. First paperback edition. 58 full-page illustrations. 512pp. 5⅜ x 8½.   26507-2

THE PRINCIPLES OF PSYCHOLOGY, William James. Famous long course complete, unabridged. Stream of thought, time perception, memory, experimental methods; great work decades ahead of its time. 94 figures. 1,391pp. 5⅜ x 8½. 2-vol. set.
Vol. I: 20381-6   Vol. II: 20382-4

THE WORLD AS WILL AND REPRESENTATION, Arthur Schopenhauer. Definitive English translation of Schopenhauer's life work, correcting more than 1,000 errors, omissions in earlier translations. Translated by E. F. J. Payne. Total of 1,269pp. 5⅜ x 8½. 2-vol. set.   Vol. 1: 21761-2   Vol. 2: 21762-0

MAGIC AND MYSTERY IN TIBET, Madame Alexandra David-Neel. Experiences among lamas, magicians, sages, sorcerers, Bonpa wizards. A true psychic discovery. 32 illustrations. 321pp. 5⅜ x 8½. (Available in U.S. only.)   22682-4

THE EGYPTIAN BOOK OF THE DEAD, E. A. Wallis Budge. Complete reproduction of Ani's papyrus, finest ever found. Full hieroglyphic text, interlinear transliteration, word-for-word translation, smooth translation. 533pp. 6½ x 9¼.   21866-X

MATHEMATICS FOR THE NONMATHEMATICIAN, Morris Kline. Detailed, college-level treatment of mathematics in cultural and historical context, with numerous exercises. Recommended Reading Lists. Tables. Numerous figures. 641pp. 5⅜ x 8½.
24823-2

PROBABILISTIC METHODS IN THE THEORY OF STRUCTURES, Isaac Elishakoff. Well-written introduction covers the elements of the theory of probability from two or more random variables, the reliability of such multivariable structures, the theory of random function, Monte Carlo methods of treating problems incapable of exact solution, and more. Examples. 502pp. 5⅜ x 8½.   40691-1

THE RIME OF THE ANCIENT MARINER, Gustave Doré, S. T. Coleridge. Doré's finest work; 34 plates capture moods, subtleties of poem. Flawless full-size reproductions printed on facing pages with authoritative text of poem. "Beautiful. Simply beautiful."–*Publisher's Weekly.* 77pp. 9¼ x 12.   22305-1

NORTH AMERICAN INDIAN DESIGNS FOR ARTISTS AND CRAFTSPEOPLE, Eva Wilson. Over 360 authentic copyright-free designs adapted from Navajo blankets, Hopi pottery, Sioux buffalo hides, more. Geometrics, symbolic figures, plant and animal motifs, etc. 128pp. 8⅜ x 11. (Not for sale in the United Kingdom.)   25341-4

SCULPTURE: Principles and Practice, Louis Slobodkin. Step-by-step approach to clay, plaster, metals, stone; classical and modern. 253 drawings, photos. 255pp. 8⅛ x 11.
22960-2

THE INFLUENCE OF SEA POWER UPON HISTORY, 1660–1783, A. T. Mahan. Influential classic of naval history and tactics still used as text in war colleges. First paperback edition. 4 maps. 24 battle plans. 640pp. 5⅜ x 8½.   25509-3

CATALOG OF DOVER BOOKS

THE STORY OF THE TITANIC AS TOLD BY ITS SURVIVORS, Jack Winocour (ed.). What it was really like. Panic, despair, shocking inefficiency, and a little heroism. More thrilling than any fictional account. 26 illustrations. 320pp. 5⅜ x 8½.
20610-6

FAIRY AND FOLK TALES OF THE IRISH PEASANTRY, William Butler Yeats (ed.). Treasury of 64 tales from the twilight world of Celtic myth and legend: "The Soul Cages," "The Kildare Pooka," "King O'Toole and his Goose," many more. Introduction and Notes by W. B. Yeats. 352pp. 5⅜ x 8½.
26941-8

BUDDHIST MAHAYANA TEXTS, E. B. Cowell and others (eds.). Superb, accurate translations of basic documents in Mahayana Buddhism, highly important in history of religions. The Buddha-karita of Asvaghosha, Larger Sukhavativyuha, more. 448pp. 5⅜ x 8½.
25552-2

ONE TWO THREE . . . INFINITY: Facts and Speculations of Science, George Gamow. Great physicist's fascinating, readable overview of contemporary science: number theory, relativity, fourth dimension, entropy, genes, atomic structure, much more. 128 illustrations. Index. 352pp. 5⅜ x 8½.
25664-2

EXPERIMENTATION AND MEASUREMENT, W. J. Youden. Introductory manual explains laws of measurement in simple terms and offers tips for achieving accuracy and minimizing errors. Mathematics of measurement, use of instruments, experimenting with machines. 1994 edition. Foreword. Preface. Introduction. Epilogue. Selected Readings. Glossary. Index. Tables and figures. 128pp. 5⅜ x 8½. 40451-X

DALÍ ON MODERN ART: The Cuckolds of Antiquated Modern Art, Salvador Dalí. Influential painter skewers modern art and its practitioners. Outrageous evaluations of Picasso, Cézanne, Turner, more. 15 renderings of paintings discussed. 44 calligraphic decorations by Dalí. 96pp. 5⅜ x 8½. (Available in U.S. only.) 29220-7

ANTIQUE PLAYING CARDS: A Pictorial History, Henry René D'Allemagne. Over 900 elaborate, decorative images from rare playing cards (14th–20th centuries): Bacchus, death, dancing dogs, hunting scenes, royal coats of arms, players cheating, much more. 96pp. 9¼ x 12¼. 29265-7

MAKING FURNITURE MASTERPIECES: 30 Projects with Measured Drawings, Franklin H. Gottshall. Step-by-step instructions, illustrations for constructing handsome, useful pieces, among them a Sheraton desk, Chippendale chair, Spanish desk, Queen Anne table and a William and Mary dressing mirror. 224pp. 8⅛ x 11¼.
29338-6

THE FOSSIL BOOK: A Record of Prehistoric Life, Patricia V. Rich et al. Profusely illustrated definitive guide covers everything from single-celled organisms and dinosaurs to birds and mammals and the interplay between climate and man. Over 1,500 illustrations. 760pp. 7½ x 10⅛. 29371-8